LYALL WATSON

Jacobson's Organ

and the Remarkable
Nature of Smell

PENGUIN BOOKS

PENGUIN BOOKS

Published by the Penguin Group
Penguin Books Ltd, 27 Wrights Lane, London w8 5TZ, England
Penguin Putnam Inc., 375 Hudson Street, New York, New York 10014, USA
Penguin Books Australia Ltd, Ringwood, Victoria, Australia
Penguin Books Canada Ltd, 10 Alcorn Avenue, Toronto, Ontario, Canada M4V 3B2
Penguin Books India (P) Ltd, 11, Community Centre,
Panchsheel Park, New Delhi – 110 017, India
Penguin Books (NZ) Ltd, Private Bag 102902, NSMC Auckland, New Zealand
Penguin Books (South Africa) (Pty) Ltd, 5 Watkins Street,
Denver Ext 4, Johannesburg 2094, South Africa

Penguin Books Ltd, Registered Offices: Harmondsworth, Middlesex, England

First published by Allen Lane The Penguin Press 1999
Published in Penguin Books 2000
1

Printed in England by Clays Ltd, St Ives plc

The most mysterious, the most human thing, is smell . . .
COCO CHANEL, in *Her Life, Her Scents*

Contents

vii

CONTENTS

Right under our noses

We have a nose for things. We sniff out problems and follow our noses, moving to a successful conclusion in directions that are often far from obvious. And yet we persist in belittling our olfactory abilities, dismissing the human nose as a blunt instrument.

It is true that most animals have a more acute sense of smell. Dogs are a million times more likely to pick up social scents, hedgehogs ten thousand times better equipped for finding food. But, even with olfactory centres that occupy just one-thousandth of our cranial capacity, we humans are remarkably good at detecting, recognizing and remembering odours.

We can identify our relatives by smell alone, or follow the menstrual cycles of our friends and lovers. We can diagnose disease, detect danger, and distinguish between good and bad food just with our noses. We know, along with Rousseau, that 'Smell is the sense of memory and desire.' We realize, with Diderot, that smell is the most voluptuous sense. And we acknowledge the truth in Helen Keller's contention that smell, while representing the 'fallen angel' of the senses, nevertheless remains a 'potent wizard that transports us across thousands of miles and all the years we have lived'.[95]*

So why the ambivalence? It seems we are missing something. I

---

* Numbers in the text (e.g. [95]) refer to sources listed numerically in the Bibliography.

think we are, and I believe I know what it is. It is an obscure body part. One that has been there, right under our noses, all the time. Science has known about it since 1811. Biologists are familiar with it as the structure in the roof of a snake's mouth which 'tastes' molecules collected by the reptile's flickering tongue. And a few dedicated anatomists have tracked down something similar in the nasal passages of possums, anteaters, bats, cats, rabbits, and even a white whale. But, though this body part was also discovered and described in humans a century ago, it has since mysteriously disappeared from the textbooks.

There are passing mentions of it in medical and technical journals dedicated to comparative anatomy and olfactory physiology, where it tends to be dismissed as something vestigial, an anatomical ghost that makes a transitory appearance in the human embryo, vanishing well before birth. And yet it still exists: a study in 1991 of a thousand randomly selected adult human subjects found it in the noses of almost every one.

This elusive feature is the Organ of Jacobson, named after the sharp-eyed Danish anatomist who discovered it nearly two centuries ago. It is easy to miss. The external evidence consists simply of a pair of tiny pits, one on either side of the nasal septum, a centimetre and a half above every human nostril. But the fact that it does exist changes everything. With it, we reinherit the possibility of a powerful and ancient chemical sense: an ability to enter once more into a system of subliminal signalling that continues to give other animals access to a world we thought we had lost as a result of our emphasis on sight instead of smell.

Jacobson's Organ rescues our most underrated sense from obscurity. But it is not simply a supercharger, making us more sensitive to odours. What it seems to do is to open up a channel quite separate from the main olfactory system. It feeds an older, more primal area of the brain, one that monitors airborne hormones and a host of other undercover patterns of information, making physiological changes that have profound effects on our awareness, on our emotional states and on our most basic behaviours. Recent research suggests that this system could be the mechanism necessary for operating a true 'sixth sense', one that may account for our sometimes apparently supernatural ability to receive information not normally available to the traditional five faculties.

If all this is true, Jacobson's Organ could be the most important key to unravelling the mysteries of our minds since the discovery of the unconscious. As an evolutionary biologist and anthropologist, that possibility excites me enormously.

Aristotle associated the four fundamental elements of earth, air, fire and water with the four basic senses of sight, hearing, touch and taste. But he also recognized a fifth element, the most essential substance of all, called *quintessence* – and connected this to the sense of smell. That, he suggested, lay at the heart of perception, linking the other four senses. Ever since then, five has been the established number of the senses in Western culture. I have no argument with Aristotle, but I believe it is time to extend his fifth sense, already at the centre of awareness, into new areas, expanding smell in ways that give our Cinderella sense its proper due.

Lyall Watson
Castlemehigan, Ireland, November 1998

PART ONE

GETTING ORGANIZED

Smell is the forgotten sense. There are no agreed measures of its nature, no societies dedicated to its appreciation, no descriptions of it except those borrowed from our overbearing sense of sight.

Smell is our most seductive and provocative sense, invading every domain of our lives, providing the single most powerful link to our distant origins. But it is also mute, almost unspeakable, defying description and collection, challenging the imagination. All that stops it slipping entirely through the net of language is a few brave attempts to pin it down – beginning with the work of a very tidy-minded Swede.

Carolus Linnaeus (1703–78) was the Great Indexer. He studied medicine at Uppsala University, but his heart was always in botany. He began his green studies with a survey of all the flowering plants of Lapland, completing this task in 1737 with the aid of a revolutionary new system of defining and describing every species. And during the next twenty years he engineered a huge expansion in natural history, publishing his *Systema naturae* and docketing everything in sight, changing for ever the way we think about the world around us.[114]

Knowing the names of things gives us power over them: the power to isolate them from nature. This separation remains artificial but also extraordinarily useful, making it possible for us to set the trees aside for a moment and concentrate instead on the composition and coexistence of the forest. Labelling is an essential first step in coming to terms with ecology, and Linnaeus, in his fever to catalogue all of existence, applied his art even to the elusive world of smells, scents and stenches. There were, he decided, seven major classes of odour, ranging from the pleasant to the unpleasant, from those he described as 'kindly and desirable to our nerves and even to life itself' to those that were patently 'repellent to life'.[115]

Linnaeus published *Odores medicamentorum* in 1752 and there have been dozens of attempts during the last two and a half centuries to refine this system, approaching it from the vantage points of psychology, chemistry, physiology and perfumery. Some of these refinements are convenient for cosmetic chemists and those working in the fragrance industry, but even the most sophisticated new taxonomies are unsatisfactory and inconsistent in the end. They founder on the accidental nature of most smells and on the lack of a specialized vocabulary for the act of smelling in any language.[115]

There is no semantic tradition, no critical study of the origin and function of words used to describe smells in any country, and no learning process in any culture assigned specifically to the sense of smell. So I find myself returning, time and smell again, to the classification provided by the man whom novelist John Fowles calls 'The Great Warehouse Clerk of Nature'.[57] And finding solace, even a surprising new depth of meaning, in the resonant succession of Fragrantes (fragrant), Hircinus (goaty), Ambrosiacos (ambrosial), Tetros (foul), Nauseosos (nauseating), Aromaticos (aromatic) and Alliaceos (garlicky).

Linnaeus was far more than just a pigeon-holer. He was a master organizer, sorting out and tidying up the unruly strands of life in all its guises. But he was also wonderfully intuitive at it, coining names and creating patterns which for the most part still work, and which help to reveal real relationships and affinities where none are immediately obvious. He enriched sight with insight and gave us tantalizing glimpses of a grand design. It was he, of course, who gave us a place in the natural world. He allowed us to be *sapient* – which refers to someone who is not necessarily 'wise', but more properly and modestly 'would-be-wise' – perhaps by coming to our senses.

In the Andaman Islands, the Ongee people consider smell not as an isolated sense, but as a fundamental cosmic principle. For them, odour is the source of personal identity. It produces life and causes death. When Ongee – or anyone in modern Japan – refer to 'me', they put their forefinger on the tip of their nose. That is where the spirit resides, and too much or too little of it can cause problems. A healthy person is one who has their smell 'tied tightly'. Losing your odour entirely can kill you.[29]

The ideas of life and breath and spirit and smell are intertwined in many cultures. Some Mexicans still believe that the smell of a man's breath is more responsible for conception than his semen is. And in the Andamans, they tie everything very tidily together in a tradition of communication by smell which they call *mineya-lange*, which literally means 'to remember'.

Nothing is more memorable than a smell. Ask anyone to recall a childhood home or friend, and the details are likely to be sketchy. But provide just one whiff of a familiar scene, and the memories come rushing back, not piece by piece, but as a whole, with all the flavours of the original experience miraculously intact. And the effect is explosive. As Diane Ackerman says, 'a complex vision leaps out of the undergrowth'.[1]

Indeed it does. Perhaps this is because smell is the only sense you can't turn off. You can close your eyes, cover your ears, refrain from touching or tasting. But we smell all the time and with every breath, twenty thousand times a day. And if I am right about Jacobson's Organ, all that information is filed away somewhere – not in the grey matter of the conscious brain, which is far too busy with things of the moment, but in the warehouses of long-term memory, in those parts of the brain more touched by sensibility than mere sensation. There is a big difference between sense and sensibility. And because my exploration of a possible sixth sense will take in several kinds of olfactory experience, it is important here, at the outset, to know something about how smell works.

Noses are conspicuous. They hold centre stage on our faces, reaching out into the world, demanding attention. We don't count mouths, we count noses. We look down them, pay through them and run the risk of having them put out of joint.

We breathe through our noses, warming incoming air, but any simple opening would have sufficed for that. What we have instead is an exaggerated profile involving a canopy supported by a clearly purposeful piece of cartilage. This keeps the rain out, deflects water when we swim, and may even add resonance to our speech, but functionally it has most in common with a ventilator hood – something like an air scoop fitted to the deck of a boat. The nose projects away from the olfactory confusion of the face, avoiding self-smelling, inhaling the rest of the news. Testing the air for early warnings gives those who stand guard, usually the men, noticeably larger nasal structures.

All our noses are alike inside, opening into a pair of cavities separated by a septum. These are surprisingly large vaults, occupying almost as much space in the skull as our famously big brains. Most of this is air-conditioning apparatus, divided up into three horizontal chambers by thin, scrolled bones covered in vascular tissue which expands and shrinks in tune with a variety of reasons and seasons. So the flow of air through the left and right nostrils is seldom equal, giving us alternating rather than parallel passages, and creating conditions which are always turbulent. Perhaps, again, with good reason.

High on the roof of the two upper chambers, roughly on a level with our eyebrows, there is a patch of yellowish tissue in a pair of clefts. Each patch is just one centimetre square. Put together, the pair would fit on a postage stamp, but both are richly endowed with several million fringed receptor cells. This is where smells are collected.

The position of these olfactory epithelia, hidden away in the back of the nose, seems awkward. Logically, they ought to be out there where the action is, bathed in a constant stream of air. But like the other sense systems, the olfactory organ needs variety. Just as spontaneous tiny eye movements provide an essential variation of stimulus to the retina, so the nose thrives on subtlety and change. It lies in wait for a whiff of something new and interesting, and then it demands attention – and we wrinkle up our noses and sniff.

Smell is a chemical sense. What the receptor cells in the nose do is translate chemical information into electrical signals. These travel along olfactory nerves into the cranial cavity, where they gather in the olfactory bulbs. These, in turn, feed the cerebral cortex, where association takes place and nameless signals become transformed into the fragrance of a favourite rose or the musky warning of an irritable skunk.[5]

Trained noses can identify hundreds of thousands of odours, far more than our minds can describe. Smells sit right on the tops of our tongues and linger there, because we cannot utter their names. It is almost impossible to describe even the most familiar scents to someone else who has never smelled them. And part of this confusion exists, I believe, because of the way we ourselves experience them.

The odour-to-nose-to-brain sequence I have just described in very bald terms is *not* the only way we smell. Our nasal chambers are connected to the outside air directly through the nostrils, and indirectly through the throat by 'inside air' that comes perfused with the flavours of food and drink, and the products of digestion. All this information goes to the olfactory bulbs of the brain. But riding on the side of each of these swellings in our heads are other smaller lobes called the accessory bulbs, which get their news of the world from an altogether different source: the twin-tubed Organ of Jacobson.

Twenty years ago, hardly anyone believed in such a thing. There were old reports from Victorian anatomists, but the modern anatomical consensus was that humans don't have such an organ. All that changed at the University of Colorado in 1991. Bruce Jafek was a practising surgeon then, specializing in nose jobs, until he became curious about Jacobson's Organ. He and microscopist David Moran devised a simple nasal speculum to help in the search, and by laying people on their back and shining a light up their nose, they discovered the organ in every one of the two hundred people they examined.[129]

Once you know exactly where to look, finding it isn't difficult.

The openings can usually be seen with the naked eye as tiny, pale pits, near the front of the nasal septum, about a centimetre and a half in from each nostril, just above the floor of the nose. Sometimes the pits are relatively large, a couple of millimetres across; sometimes a binocular microscope is necessary to find them. But everyone, regardless of age, sex or race, has a pair – unless you have had the sort of cosmetic surgery which removes that part of the nasal septum entirely.

These pits lead to two short tubes equipped with sensory cells quite different from those in normal olfactory tissue. Moran said, 'They don't look like any other nerve cells I have ever seen before in the human body.'[214] But in 1990, physiologists Luis Monti-Bloch and Larry Stensaas, of the University of Utah, examined a further four hundred subjects and found that all of them had such twin tubes, and every one they tested was indeed using these organs to send messages to the brain.[128]

But the most interesting thing is that Jacobson's Organs are not receptive to ordinary odours. They respond most often to a range of substances which have large molecules, but often no detectable odour. And they communicate not with the olfactory bulbs and the cortex, but with the accessory bulbs and that part of the brain that coordinates mating and other basic emotions. Recent evidence suggests also that the two separate and parallel systems of odour detection cooperate in surprising ways to produce novel sensibilities not achievable by either of them on their own.

Moran, Jafek, Stensaas and Monti-Bloch are now part of a small group of distinguished scientists who believe they have discovered a new sense organ, one that detects chemical signals previously thought to be beyond the scope of human sensitivity. They call it the *vomeronasal organ*, rather than Jacobson's Organ. But by any name it has put them at odds with conservative physiologists and neurologists who want more concrete evidence, preferably from humans willing to have dyes injected into their systems, so that their brains can later be scanned to see where the tracers end up. Such foot-dragging in the face of facts has a familiar feel to it. The last time I saw anything like it was in the 1960s, when diehard geologists made last-ditch stands against the tide of continental drift.

I am impressed by the evidence we already have for an organ I continue to credit to its Danish discoverer, and I am fascinated by the potential it has for all students of the unusual. The Organ of Jacobson seems to feed the primitive brain. It is not our olfactory link to consciousness, but rather a chemical clearing house for subliminal impressions, for all the things that lead to what science writer Karen Wright calls 'bad vibes, warm fuzzies, instant dislikes and irresistible attractions'.[214] Just the sort of will-o'-the-wisps with which any decent sixth sense ought to be concerned.

But before exploring the stranger fringes of the sense of smell with this in mind, I want to do what a good evolutionary biologist should, and look back, with the help of Linnaeus, on how it all began.

Fragrantes

For this group of odours, Linnaeus specified floral sources such as jasmine, saffron and wild lime – all of which are distinctly perfumed.

But there is something in his description which suggests that, as a Latin scholar, he was aware of possible confusion between the verbal roots fragrare ('to smell'), and flagrare ('to burn') and felt that both were appropriate.

Saffron is not just fragrant, but also fiery, glorious, even gaudy, giving Alexander Pope good reason to describe the fleet in which Odysseus set sail as 'a cruise of fragrance, formed of burnished gold'.

Saffron is a crocus. Crocus is the Chaldean name for the iris family. And Iris is the Greek goddess of the rainbow and messenger of the gods.

There is more than enough biological and mythological coincidence in all this to keep a classical scholar happy and to augur well for the evolution of a sense able to appreciate something both bright and beautiful.

I

A nose is born

Smell was our first sense. It is even possible that being able to smell was the stimulus that took a primitive fish and turned a small lump of olfactory tissue on its nerve cord into a brain. We think *because* we smelled.

The argument is simple. Before sight and sound hijacked our attention, we shared with all life a sort of common sense, a chemical sense that depended on direct contact with matter in the water or the air. And for 90 per cent of our time on Earth, that was how things worked.

Then the emphasis shifted. We learned to live instead with waves of energy, making sense of chaos, and becoming conscious. Which, of course, was a good thing . . . but we need now to go back and pick up some useful talents that we abandoned on the way.

Four hundred million years ago, in the Devonian period, fish were the most important and advanced animals on Earth. And the most abundant of them were jawless, heavily armoured forms, filtering their food out of thick coastal muds. Theirs was a world of direct sensation. Behaviour was simple. Any unusual stimulus, a bright light, a loud sound, any abrupt contact, produced much the same response: one of aversion. They pulled back and tried again later, taking life one taste at a time.

In the beginning, it was difficult to separate taste and smell. In the mud, sensation followed contact. You bumped into things and then tested them to see whether they were worth avoiding or eating. The tests were chemical ones, made by cells designed to analyse molecules dissolved in the water.

Many crustaceans still work this way, using cells on their legs which respond only to amino acids, providing a ready distinction between organic and inorganic matter. Filter-feeding fish didn't need to be much more discriminating, extracting what was useful from the mud and rejecting the rest. But instead of just oozing around everywhere with their jawless mouths wide open, it would have helped these mindless pioneers to have some way of finding the most nutritious muds. What they needed was the ability to test and taste from afar – and that is where smell comes in.

Smell is a long-distance sense, a way of stretching time and finding out in advance what lies ahead. It expands both awareness and opportunity, providing a need for analysis that normally never touches the lives of bottom feeders. But it was, in fact, one of those mudsuckers that made the big leap. And the result is visible still, frozen in the life histories of a few modern species of jawless, limbless, boneless and often blind hagfish and lampreys.

Hagfish are patient scavengers, spending most of their lives embedded in soft mud with only their blunt snouts protruding, waiting for the same chemical signal that attracts crustaceans to putrid fish. But these primitive fish have an advantage over lobsters and shrimps. They have a single opening above the mouth that leads to a paired chamber in which scents can be isolated, analysed and perhaps even located. They have, it seems, the world's first noses.

Armed with these secret weapons, hagfish have flourished. At least twenty species survive, some so well that they have become a nuisance to modern fishermen, burrowing into catches of haddock and cod and consuming netted fish from the inside out, turning them into bags of bones. Hagfish see nothing, but clearly smell very well, swimming up a gradient of fishy flavours, travelling with undulating movements, turning always in the direction of the strongest stimulus, choosing the right odour corridors, keeping on going until they get there.

Their relatives the lampreys have refined the process even further, responding instinctively to just one aromatic chemical that forms part of the normal body odour of live schooling fish such as trout. Sensing this, they left their muddy larval haunts, developed functional eyes, and now actively pursue and parasitize such hosts. Once within range, the swift eel-like lampreys use their rasping teeth to hold on to any soft part of a trout, sucking out its body fluids like aquatic vampires, keeping their victims' blood flowing just as bats do, with anticoagulants in their saliva.[102]

The unpleasant habits of round-mouthed hagfish and lampreys are vividly described as 'suctorial'. Theirs is a mode of life made possible by having gills which open directly into the throat, so that they can continue to respire while still sucking blood. They are sufficiently well adapted in this respect to have survived for over four hundred million years, despite their many other primitive features. But there is one important way in which they have changed: they have not only noses, but also a *nose-brain*.

The nervous system of these suckers is rudimentary. There are no sympathetic or autonomic nerves at all: none of the networks which, in more modern vertebrates, serve the intestine, the liver, all glands and the heart. But the nerves from the nose gather in a bundle in the head which is already substantial and beginning to spread out sideways, expanding in direct proportion to its employment. The most exciting feature of this growth is that, even in this early stage, it is starting to take on the shape of something new. It is a burgeoning forebrain, bursting into life in direct response to an olfactory need.

In a habitat where sight is limited and sound cannot be localized, only smell can pick up traces that provide solid information from afar. But smell on its own is not enough. There has to be a way of telling where a smell is coming from. Hagfish and lamprey put themselves into an odour corridor by swinging their bodies from side to side, sampling water on a broad front with their single nasal opening. But most later fish have two external 'nares', or nostrils, and practise 'stereo smell', which improves with the distance between the openings. The wider the head, the better the chance of smelling in stereo. And somewhere in the space between the nostrils, there was room for a coordinator: a place where information about smells could be analysed and acted upon.

Fish live in an environment where even substances that are only partially soluble find their way to chemically sensitive areas of the body, making it possible for extraordinary feats of detection. The Common European eel, for instance, has been shown to have a very uncommon affinity for some alcohols, responding to just a few molecules at a time in concentrations as low as one part in 10^{18}. This represents the sort of dilution you would get by tipping a single shot of vodka into a body of water the size of Lake Erie.[189] These eels are, of course, teetotallers. But they spawn and die in the depths of the Sargasso Sea, leaving leaflike larvae to make their own way, unguided, across five thousand kilometres of open ocean, back to ancestral haunts in the pools and lakes of Poland and Germany, in a journey that lasts three years. How they do this remains mysterious, but it seems to have something to do with smell.

Experiments with salmon show that they are able to distinguish between plain water and water in which an aquatic plant has been briefly rinsed. Young salmon probably imprint on the particular flavours captured by their home streams from the unique combination of plants that grow only in that watershed. It is difficult to imagine how young eels manage without such crucial early experience, yet they do. They are driven by instinct, and by the presence or absence of just a few special molecules, to make decisions on which the very survival of their species is at stake.[74]

Smell *is* stimulating. It stirs things up and makes us nostalgic – a wonderful word which literally means 'ache for home' – which serves to inspire new circuits in the brain. Enough, in simpler species, to form a simple nose-brain. In migratory fish such as salmon, this centre has split into a paired organ, more befitting species that have begun to perceive and to perform bilaterally. These are the *olfactory bulbs*.

Many sharks find their food by smell. If water in which live fish have been kept is siphoned into a shark tank, the residents respond with hunting behaviour. And if the stream comes from injured or dead fish, they can be provoked into a feeding frenzy. The excitement of the sharks increases directly with signs of stress from their prey. They detect and respond to the smell of trouble, and do so with extraordinary accuracy. In one experiment with a white-tip shark, an injured fish was allowed to swim the length of a tank before taking cover. When the shark was released soon afterwards, it followed exactly along the same zigzag line, duplicating every movement made by the now-invisible prey.[190]

It seems that smell leads these predators to their prey from a distance, but the final move, the kill, may be triggered by another sense. For some species, in clear water and in daylight, this is often vision. In others it may be pressure waves detected by sensory cells along the flank. But in small, bottom-living sharks such as the spotted dogfish, the focus appears to be a weak electrical field produced by another fish's muscles. And all these impressions seem to be enhanced by smell, even linked with it in some strange way.[91]

Such synesthesia, the blending of one sense with another, is vital. Even in the lives of fishes, sensation is seldom a matter of one thing or another. Senses overlap. The lines between them often tend to be blurred, and the best that we can manage, by way of description from the outside, is to say that the senses of fishes appear to dominate one at a time.

In the case of the spotted dogfish, smell slides seamlessly into an awareness which, if the fish could tell us, it might have to describe as a flavour rather than an odour. I find it fascinating that this very dogfish is renowned for having a forebrain which is not only large, but also equipped with a pair of swollen olfactory bulbs, each proportionately larger and better differentiated than those of almost any other fish. It is well prepared for blurring sensory boundaries in the same way that we, when faced with perceptual confusion, resort to metaphors, talking of 'loud smells' or 'bright sounds'.[177]

Dr Johnson experienced the colour scarlet as 'the clangour of a trumpet', and the poet Rimbaud described the sound of the vowel *a* as 'a black hairy corset of flies'. And it can be no coincidence that natural synesthetes, those who experience intense sensory crossovers on a regular basis, have problems with their limbic systems – those areas of the brain that, in mammals, grew out of the old olfactory bulbs.

One physiologist has even described human sensory blenders as 'living cognitive fossils', displaying a memory of how our early vertebrate ancestors once saw, heard, touched, tasted and smelled.[119] It is a fact, too, that our cerebral hemispheres, those walnut-folded swellings of grey matter that dominate the forebrain and now control most of our conscious behaviour, also developed directly from olfactory tissue.

The limits of sensory evolution in fish are defined very largely by their habitat. Water is physically supportive, carries some kinds of odour well, and is kind to sound – letting it travel several times faster than air will allow, but it inhibits other more personal kinds of communication. Much more becomes possible when scents are released into the air. Breathing air is a liberating experience. It freed our ancestors from the constraints of staying wet or having to remain within easy reach of water for refuge, respiration or reproduction. But the biggest change it made in our lives was to expose us to a whole new range of sensory experience.

Air is traditionally 'thin', but the more we learn about our atmosphere, the more substantial it becomes. In some places it is so filled with inorganic flotsam that it is almost thick enough to plough; in others, it has become so primed with the by-products of life that it comes close to being a living tissue in its own right. Even the cleanest air, at the centre of the South Pacific or somewhere over Antarctica, has two hundred thousand assorted bits and pieces in every lungful. And this count rises to two million or more in the thick of the Serengeti migration, or over a six-lane highway during rush hour in downtown Los Angeles.

Most airborne material consists of minute particles of salt, clay, and ash from forest fires and distant volcanic eruptions. And mixed in with, or growing on, or simply being carried along by this fertile soil is a garden of exotic flora and fauna. Every lungful of air we borrow from this gruel is likely to contain a few stray viruses in transit between their hosts; four or five common bacteria; fifty or sixty fungi, including several rusts or moulds; one or two minute algae drifting in from the coast; and possibly a fern or moss spore, or even an encysted protozoan.[202]

All of which is inevitable. This is, after all, the stuff of life. We share our planet quite naturally with a permanent aeroplankton;

a buoyant ecology too soft to hear, too small to see, but heavy with mood and meaning. Imagine being aware of all these airy inclusions – and you can begin to understand how it might feel to be able to smell really well.

Air-breathers do things differently. Their olfactory sense cells lie, not in isolated nasal sacs, as in fishes, but strategically placed in a nasal passage, a thoroughfare through which air passes on its way to the lungs. And that simple fact gives noses a new slant. The difference is most dramatically demonstrated in an intermediate animal, one that has a nose in both worlds, something like a frog or a toad that is amphibious, literally leading a double life. The African clawed toad is the white rat of this arena.

Every gynaecological laboratory has one, largely because it provided the first reliable way of testing for human pregnancy. An unfertilized female toad lays eggs within hours of being injected with urine containing hormones that are characteristic of newly pregnant women. But these smooth, streamlined, strange-footed toads – whose triple claws are used for digging for food on the bottom of South African ponds – are rapidly becoming equally popular with geneticists.

At the Institute of Zoophysiology in Stuttgart, work on the toad genome has revealed that 'Strange Foot' has an equally strange genetic repertoire. This includes hundreds of genes that code for the way in which its olfactory sense cells (called *receptor cells*) can function. That is no great surprise: mammals have thousands of such genes while fish have very few, so it was predicted that amphibians would be intermediate between the two. And so they are. Their olfactory genes fall into families of two distinct kinds.[58]

Aquatic animals are exposed to water-soluble molecules such as amino acids, while air-breathers have access to a far greater variety of more volatile odorants. And as an amphibian, adapted to both aquatic and terrestrial life, the clawed toad was expected to have roughly equal numbers of both genes and receptor cells. It doesn't. There are far more cells of the mammalian air-breathing kind, probably because there are far more airborne than water-borne odours. But the big surprise is that the two kinds of receptors are found in separate areas. Strange Foot has two noses and two senses of smell: one for use underwater, the other for life in the air.

The toad's nose is neatly divided into two separate sacs or chambers. Just inside each nostril there is a flap of tissue which acts like a valve between them. Underwater, the flap swings to seal off the main chamber, exposing another one, a cul-de-sac filled with receptors sensitive to water-borne odours. And when the toad surfaces and raises its nose into the air, the flap swings the other way to seal off the aquatic chamber, exposing the main one, which is lined with cells sensitive to volatile odours in the air. And after serving this chamber, the airstream passes on to provide oxygen to the lungs.

Living in the air demands more of the sense of smell. But the challenge seems to be one that even the most primitive amphibians are ready to meet.

Mexican toads, in the process of learning to negotiate a maze, have been trained to remember alien scents such as geraniol, vanillin and cedarwood.[70] Both the spotted chorus frog and Strecker's chorus frog have shown that they can pick up the odour of their own breeding ponds many hundreds of metres away.[69] North American leopard frogs, blinded by severing their optic nerves behind the eye, can nevertheless navigate their way back to a familiar pool, whether they are released upwind or downwind of their homes.[41] But it is the olfactory skill of the Californian red-bellied newt that takes one's breath away.

This newt is lizard-like, with large dark eyes and tomato-red markings that warn of a poisonous secretion in its skin. It has few problems with predators in the coastal mountains where it lives, but has been persecuted instead by local biologists interested in its almost unbelievable navigational skills. It can be taken away from its usual damp haunts in one valley, carried across a mountain divide three hundred metres high, and still make its way unerringly back home.[193] If moved in any direction, it will succeed, even in totally strange territory, in orienting itself quickly and directly towards home.

Blinding this nostalgic little animal doesn't slow it down for a moment. And squirting formaldehyde into its nose to kill the epithelium – the thin lining of tissue that contains olfactory sense cells – results only in what has been described as 'a significant loss of orientation ability'.[67] The only way, it seems, to prevent it from getting home is to sever its olfactory nerves entirely. Such invasive studies distress me. But the fact that they were thought to be necessary is a measure of our frustration when faced with behaviour for which we have no easy explanation. Such mysteries lie in every part of natural history. Even the simplest creatures have the capacity to surprise us when we think boldly enough to ask them the right kinds of questions. And the answers are never easy, except in hindsight.

In the case of the wandering newts, however, there may be a

clue. There is something I have so far failed to mention about amphibians, something that began with those frogs and toads whose noses have divided into the two chambers appropriate to their schizophrenic lifestyle. It is this: some of them, including this newt, also have a third nasal space which is not directly involved in detecting smells, either underwater or in the air. And this could be involved in a new and different kind of awareness.

The olfactory sense cells of all vertebrates are, on the whole, surprisingly alike. And also very strange. They are, for a start, unique in that they are in direct contact with the outside world. Most other sense organs lurk under the skin, embedded deep in protective tissue, picking up sensations by remote control. But smell cells go unadorned: naked neurons, each one right out there in the open, like a unicellular organism, meeting molecules, making its own way in the world. And, strangest of all, small cells wear out after a few weeks and need to be replaced on the front lines. They regenerate in ways no other nerve cells in our bodies ever do.[184]

The number of olfactory sensory cells operating at any one time varies from one species to another, depending on the emphasis a species places on having a good sense of smell. Humans, for example, have about six million cells; rabbits fifty million; and sheep dogs over two hundred million.

In all species the cells are thin, with projections, or processes, at either end. One of these extends inside the body, usually through a hole in the skull, to make contact with the brain. The other process is far shorter and stays on the outside in a knob topped with a tassel of fine sensory hairs, called cilia. The number of cilia varies too, according to the species. Domestic cattle, for example, all have exactly twenty-seven microscopic lashes on each sensory cell. Tiny newts have fewer such hairs, but theirs are four times as long.[52]

It is generally agreed that the function of these frills is to bind odour molecules and convert their chemical signals into electrical ones for transmission to the olfactory bulb in the brain. So it is interesting to discover that, in addition to typical sensory cells in the main and side chambers of some amphibian noses, there are other cells in a deeper chamber that look very different. They are flask-shaped, with long necks. They have no cilia, but have knobs decorated with a tuft of short bristles, sometimes called microvilli, that makes them look like sensory cells with a crew-cut. And the tissue in which these bristle cells lie has none of the glands which, in normal olfactory sensory epithelium, produce the mucus that helps to trap and hold airborne chemicals.[53]

No one knows exactly what these strange cells do. I suspect that they give air-breathing animals a new sensory edge. This extra faculty may be related to the sense of smell, and might even partially replace it, but it has a different, more multi-sensory bias, and a whole new perceptual agenda that becomes truly obvious only in the first fully terrestrial vertebrates – the reptiles.

Some reptiles differ very little from amphibians in both their lifestyle and their nasal structure. Marine turtles are as water-dependent as clawed toads and have very simple nasal passages that run, almost in a straight line, from the external nostril to an internal opening in the throat. Land tortoises possess larger olfactory structures, and crocodilians have a surprisingly sophisticated series of chambers and sinuses which provide additional surfaces for the usual olfactory sense cells.

Marine snakes, tree-living lizards and chameleons are all conservative about smell, but it is among ground-living lizards and snakes, those closest to the source of many scents, that the greatest changes have taken place. And the most dramatic of these is a concentration of the mysterious bristle cells in a pair of separate chambers of their own.[139] This, finally, is Jacobson's Organ.

The discoverer of the organ, Ludwig Levin Jacobson, was born in 1783 in Copenhagen. He qualified in surgery at the age of just twenty-one and went to study in Paris with the great anatomist Baron Georges Cuvier, who believed that the habits of an animal determine its form. Cuvier also refined the taxonomic system of Linnaeus by showing how all organisms could be classified on the basis of their anatomical differences. And it was probably he who drew Jacobson's attention to the work of Frederick Ruysch, a Dutch embalmer who in 1703 described the anatomy of a number of animals, including a snake with unusual pits in its palate.[161] So Jacobson looked for, in 1809 discovered, and in 1811 published a report on the organ which still bears his name.[84]

The Organ of Jacobson remained a zoological curiosity until the late nineteenth century, when interest in it was revived by one of the most charismatic figures in Victorian science. Robert Broom was a Scot and a physician with a very curious mind. One of the last great individualists, he found fault with a place or a practice only if it offered 'too much medicine and too little natural history'.[54] In later life he became one of the world's great palaeontologists and physical anthropologists, doing pioneer work on the origins of humans in Africa, but his first enthusiasm was for reptiles, their origins and the significance of Jacobson's Organ. In between practising medicine he scoured museum collections and studied animals in the wilds of Australia and South Africa for material to include in a doctoral thesis, 'On the comparative anatomy of Jacobson's Organ', which he submitted to Glasgow University.

'It would seem,' said Broom, 'that the Organ of Jacobson is the organ in the body that is least liable to become altered by change of habit. I can almost identify an animal by examining this organ and often tell of its affinities.'[16] He went on to do just that for the mammal-like reptiles that once roamed South Africa, most of which he found himself and modestly described as 'the most important fossil animals ever discovered' because 'there is little doubt that among them we have the ancestors of mammals and the remote ancestors of human beings'.[17]

Recent research confirms Broom's suspicions about Jacobson's Organ. Signs of it appear, like portents, in the embryos of all higher land mammals, although in some, such as tree-living lizards, aquatic turtles and nearly all birds, it never actually materializes. There is no need for it. In others, such as ground-living lizards and most snakes, the organ comes into being, not simply as pouches off the nasal cavity, but as a pair of blind chambers that open directly into the mouth.[47] Why should this be? What happened to make this development not just useful, but necessary?

Robert Broom, who could put flesh on the bones of a fossil even while it was still in the ground, described minute channels on the skulls of some of his finds in the fossil beds of the Karroo Desert,[78] marks which suggest that Jacobson's Organ arrived on the scene during the Triassic, perhaps two hundred million years ago. It appeared, he believed, to meet a need that could best be understood by looking at lizards or snakes that live today in similar habitats.[18]

The most obvious sensory feature of all living snakes is that they have two olfactory systems. In one, external nostrils lead to nasal cavities containing sensory cells with cilia, embedded in a tissue containing glands. The other consists of cells without cilia, in a tissue without glands or mucus, housed in a pair of dome-shaped structures on either side of the nasal septum. Access to this second system is restricted to a pair of tiny pits in the roof of the mouth, leading to ducts that pass through the bony palate.

The first system is the main olfactory apparatus, often simply known as the organ of smell, though neither of those descriptions is totally accurate. It is not the only organ, nor even necessarily the main one, involved in smell. The second is Jacobson's Organ, comprising a rival chemical sense system that is becoming the focus of very active research and debate, much of it based on one common little snake that has been studied in the field by David Crews and his colleagues at the University of Texas.[37]

Garter snakes are the most widely distributed reptiles in North America. They can be found at all altitudes, ranging from coast to coast and from Canada all the way south to Costa Rica. They are slender, graceful snakes, seldom more than sixty centimetres long, marked with dark spots and longitudinal red stripes, giving them some resemblance to those fancy coloured garters that once supported every gentleman's socks. But their appeal to biologists in the last two decades is more sensory than sartorial.

Plains garter snakes in cool areas have the habit of gathering, ten thousand at a time, in dens established in underground caverns where they spend the long winters in hibernation. At these times their blood becomes as thick as mayonnaise, and they barely move for up to six months. But in May, as soon as the outside temperature reaches 25°C, dormancy ends and the tangle of bodies stirs and turns into a writhing mass that flows out of the den.

First the males emerge en masse and wait near the entrance, sunning themselves, showing no interest in food or drink. Then the slightly larger females come out, one at a time, to find themselves heavily outnumbered and soon entangled in a mating ball of perhaps a hundred hopeful suitors. Fifteen minutes later, carrying a load of sperm in her oviducts, each solitary female sets off on a migration to summer feeding grounds, where she will restore her body fats and give birth to up to thirty live young garters in the autumn. And by late September, she and thousands of her kind gather once again in the familiar scent of some shelter or hibernaculum.[36]

This busy cycle of mating, migration, feeding, gestating and returning to hibernation, spans little more than the three summer months and presents a number of zoological problems. Some are easily resolved. The sperm each female receives in spring is stored in her oviduct for up to eight weeks, until her follicles are ready to be fertilized in midsummer. The male's reproductive organs also build up post-nuptially, expanding rapidly during the summer, when the living is easy, and he goes into hibernation packed with sperm, ready for the following spring, when the population is concentrated once more and mating is most likely to succeed. Other questions raised by the behaviour of garter snakes took longer to answer.

John Kubie and Mimi Halpern at the State University of New York have spent twenty years setting up a series of eloquent experiments designed to find out exactly how garter snakes feed and communicate.[106] They began with the common observation that snakes follow odour trails to locate their prey. Newborn young snakes of several species have been found to follow the scent of the usual prey of their species without any experience or instruction. Their ability and willingness to do so seems to be under genetic control. And when exposed to such a scent, their immediate response is to put their heads down close to the ground, flick their tongues out to actually touch the trace, and to go on doing so more and more rapidly as the stimulus grows in strength.

Kubie and Halpern set out first to establish whether a snake in such a situation uses its nose or its mouth, its olfactory epithelium or its Jacobson's Organ. They trained plains garter snakes to follow an extract-of-earthworm trail through a maze ending with the reward of an actual worm. Then they cut the nerves leading from Jacobson's Organ to the brain of some of their trainees, and found that the snakes failed all the tracking tests dismally. Their ability to follow a scent was totally disabled, while control snakes that went through the same surgical procedure, but without having the nerves cut, performed as well as they had before the operation.[108]

In a second experiment, the same researchers severed the nerves from the olfactory sense organ in the garter snake's nose, and found that this operation had absolutely no effect. These snakes followed the trail and fed as successfully as they had before the surgery, and there was no sign of any ill effects or compensation for the loss of the nasal system.

Garter snakes without their noses carry on feeding as though nothing has changed. It seems clear that snakes can follow and find their prey without their nasal apparatus, but not without their Jacobson's Organ.

While searching for a scent, most snakes are exploratory, moving slowly, swinging their heads from side to side, taking the long view, flicking their tongues out only once in a while. But once a scent has been found, the tempo changes. The snake then moves more quickly, keeping its head close to the ground, touching the surface with its tongue on almost every quick flick. There is a perceptible gear change from one mode to another, a shift perhaps from reliance on the nose and smell, to the way things 'taste'.

This shift may also represent a move from volatile scents that blow by to heavier non-volatile scents, the sensing of which depends on actual contact. It is even possible that Jacobson's Organ may have evolved specifically to deal with the sort of large molecules and heavy odours that the host cannot handle. This is an important idea. It suggests that smell operates at two levels, that it involves two parallel systems with different functions, different receptor systems and different sites in the brain. There is certainly a growing body of neurological evidence to support such a dichotomy.[153]

In all vertebrates so far studied, the olfactory nerves that run from the nose to the brain go to bundles in the olfactory bulbs, while those that leave Jacobson's Organ travel instead to the accessory bulbs. The fact that information collected by Jacobson's Organ in the garter snake does go to the accessory bulbs, has been nicely demonstrated by attaching electrodes to its head. Every time the tongue is flicked at a trail containing prey scent, an electrical charge can be recorded in the bulbs.[124]

If the main bulb is the 'nose' brain, then the accessory structure, in snakes at least, is a sort of 'face' brain, collecting information that lies somewhere between taste and smell. Beyond the olfactory bulbs, nose news goes to the olfactory cortex for assessment; while face feelings move to another neural swelling called the *nucleus sphenicus*, about which we know very little. It seems, however, to be analogous to that part of the mammalian brain in which old impulses may be integrated with more recent experience.

In this light, Jacobson's Organ begins to look like an unconscious partner to the nose. It deals with the hypothalamus rather than the more modern thalamus. It is in touch with what neurophysiologist Paul Maclean at the National Institute of Mental Health has called the 'reptilian brain', rather than the 'mammalian brain'.[117]

Jacobson is the name of the brain's olfactory autopilot.

Reptiles have relatively simple brains. In most of them Jacobson's Organ appears to provide all the information a cold-blooded animal needs to go about the daily business of finding its way about and feeding. The organ also seems to play a vital role in successful mating.[107] In the orgies of spring, male snakes approach an emerging female and explore her with tongue flicks before rubbing their lower jaws all along her back and sides in fervent courtship. Something they find there excites them, despite the fact that the skin of garter snakes has no known glands. And as soon as one of the male snakes succeeds in mating, the other suitors immediately disperse and leave the pair to themselves. None of this happens for snakes without working Jacobson's Organs.

Careful research on intact animals has shown that the male interest is sparked by a fatty substance that females produce as a precursor to egg yolk. This substance seeps from the bloodstream into the skin, and is released from gaps between the scales that open when females breathe deeply during courtship. And the substance that switches off the interest of rival males turns out to be produced by the kidneys of the mating male and to be abundant in the gelatinous plug he leaves in her cloaca. This not only acts as a mechanical obstruction to further mating, but it makes the mated female positively unattractive to other males, even rendering impotent those who happen to come into close contact with her.[71]

Mimi Halpern and John Kubie conclude that garter snakes 'depend on a functional vomeronasal system, but not on a functional main olfactory system'.[72] The simple fact is that no snake behaviour has yet been discovered which depends in any critical way on the nose.

This is a startling finding. No naturalist has ever doubted that snakes and lizards have a sense of smell. You only have to watch one explore its surroundings, flicking its tongue out to test the ground ahead, speeding up the response as it encounters anything of interest, to be certain that this is an olfactory process. Western whiptail lizards shoot their tongues out to test the 700 times an hour, which represents a serious expenditure of energy – something not to do without good reason.[173] Smell is reason enough, and the fork-tipped tongues with matching pits in the palate are clear evidence of an oral route to chemical sensation. But none of us guessed that this was the main organ of smell, that the nose didn't really matter any more.

It looks as though ancestral reptiles were faced with a choice. On emerging from the water, they already had two-chambered noses. They also had good tongues. And as far as chemical sensitivity is concerned, evolution seems to have been equivocal. The nasal route was followed, up to a point. But for ground-living species at least, the existence of a highly elastic tongue provided an attractive alternative, and soon hijacked the rival bristle-cells in their nasal pouch, and re-routed their ducts to the mouth.

Today there is a clear division of olfactory solutions among reptiles. Aquatic turtles stuck with the simple amphibian solution. Land tortoises have slightly larger and better developed noses, but also show some development of Jacobson's Organ. And ground-living snakes and lizards have gone all the way with Jacobson, to the apparent exclusion of the nose. Crocodiles are complex, showing both systems at an equivalent level of development, as befits a group that spent time on land before becoming secondarily aquatic.

But the most interesting adaptations show up in tree-living snakes and lizards. Up off the ground, smell becomes less useful, and heavy odours and most contact scents disappear altogether, and with them the need for Jacobson's Organ. Chameleons don't have one. And neither, it seems, do birds. So why do we?

Hircinos

Orchids are pollinated almost entirely by insects, specializing very often in just one species which they attract by a customized scent. Some moth orchids hedge their bets, offering lily-of-the-valley by day and attar-of-roses after dark.

For reasons which remain mysterious, many orchid scents are attractive also to ourselves, and their flowers can be persuaded to bloom in greenhouses for over six months, hanging on in hope of pollination. But those orchids which associate with flies descend to their level, with carrion calls and the stink of rotting meat.

Linnaeus singled out such species as sources of the scents he called 'hircine', from the Latin hircus for 'he-goat'. He allied the rancid odours of cheese, sweat and urine with the lifestyle of the hardy animals who still have a reputation for wanton and licentious behaviour. 'The lust of the goat,' said William Blake, 'is the bounty of God.'

Hence the goat-footed figure of Pan, symbol of fertility and transformation, the hot-blooded patron of wild nature – and the shaggy source of widespread irrational behaviour or panic.

2

In warm blood

Being warm-blooded is very expensive. It takes a lot of food and energy. An active shrew has to eat incessantly just to stay alive, but there are compensations.

Having warm blood, being able to keep body temperatures within narrow limits, has made birds and mammals the most widespread and adaptable of all terrestrial vertebrate animals. And it has given both of these classes new noses. These they need for air conditioning, for warming outside air to body temperature before it reaches the lungs. With the result that the nasal cavities of birds and mammals are divided into three separate compartments, all scrolled and folded to increase their surface area.

In the first of them, incoming air is humidified by moist tissue. In the second, the air is warmed by networks of blood vessels. And finally, saturated at something close to 32°C, the air is primed to deposit any molecules it may contain on the rich carpet of sensory cells that lines the third chamber. This is where real smelling begins . . .

In 1827 the watercolourist John James Audubon published notes on the turkey vulture in which he described a test to show that it found its food by sight, not smell.[6] This started a long-running controversy on whether or not birds have a sense of smell. The consensus was that, by and large, they did not. It was conceded by some that the bird the Greeks called *kathartes* – the 'cleanser' or 'purifier', the one that cleared up rotting carcasses – was, at best, scent-impaired.[207] But the debate continued.

Birds are feathered reptiles. The evolution of feathers from reptilian scales remains mysterious. But in almost every other way, their cold-blooded ancestry is clear, as is the drastic programme of weight reduction that made it possible for them to get airborne. When you have sacrificed your teeth and the use of your hands, it would seem simple to jettison a sense which appears to be of little relevance to your new lifestyle.

Birds, like ourselves, are eye-animals. Their investment in vision is as comprehensive as our own, and at first sight it seems to have been made at the expense of smell. Birds lack obvious noses. Nevertheless, there are persistent reports of their ability to smell, perhaps even very well, and it would be useful to know why and how this has happened, and what it might mean for other bipedal animals with their heads in the air. Like ourselves.

The turning point in the debate about birds and smell came in 1968 when neurophysiologist Betsy Bang of the Johns Hopkins University published her report on the olfactory bulb in the brain of 108 different kinds of birds. She measured the bulbs and, to give some idea of their size relative to the rest of the brain, expressed her results in terms of what she called an 'olfactory ratio'. This is the ratio of the diameter of the bulb to the diameter of the most prominent part of the same bird's brain, expressed as a percentage.[10]

The results are fascinating. The highest percentages, which presumably reflect the importance of smell to that kind of bird, belong to some of the tube-nosed relatives of the albatross.[81] And the odour-leader of the flock is the Antarctic snow petrel, with a score of 37 per cent. Perhaps this ice-white scavenger of the blizzards and glaciers needs all the help it can get to find every scrap of food in the freezer.

Just behind it, and a fraction further away from the polar regions, come Wilson's petrel and the greater shearwater, with scores of 33 and 30 per cent. Coincidentally, these two species are the very ones that turned up a few years later in the Bay of Fundy, on Canada's eastern seaboard, when zoologist Thomas Grubb dragged a pair of sponges on floats behind his boat – one soaked in cod liver oil, the other dampened only in sea water. No birds came to the unscented control sponges, but both petrels and shearwaters were very curious about the oil-soaked baits, following the airborne odour to its source in all weathers, by day and night.[68]

Unhappily for Audubon, the next highest score on the Bang scale was 29 per cent for the turkey vulture. After a hundred and sixty years, refutation of the idea that vultures can't smell arrived from Barro Colorado Island in Panama, where ornithologist David Houston sacrificed chickens and staked them out beneath the forest canopy, either hidden under vegetation or in full view. Vultures discovered few of these carcasses until they were at least one tropical day old and had begun to decay. Then they found the hidden baits as quickly as the visible ones, and showed a distinct preference for those that had decomposed the least. Houston suggests that these New World vultures not only hunt by smell, but can distinguish the same airborne bacterial toxins that lead insect scavengers to carcasses in different stages of putrefaction.[79]

Low scorers, meaning birds with olfactory ratios of less than 10 per cent, who seem to invest very little energy in smell, are mostly small perching birds such as sparrows and canaries that peck mindlessly at anything that looks like a seed.[10] The percentages tell their own story, but it would be premature to write off any species on the basis of its score alone. There are surprises in store.

Betsy Bang's work has done wonders for bird noses. In 1971 she produced her *coup de nez*, publishing detailed descriptions of the nasal apparatus of 151 species of birds in 23 different orders.[11] This leaves little doubt that most birds have all the anatomical equipment necessary to detect and track down odours. Follow-up studies have shown that they also have surprising powers of discrimination.

Birds that reuse old nest sites, for instance, have trouble with parasites that make life miserable for new chicks, but some have come up with an ingenious solution. They line their nests each season with fresh vegetation taken from plants that are known to have antibiotic and pesticidal properties. European starlings in Ohio, for instance, show a preference for small-flowered agrimony, elm-leafed goldenrod and yarrow – despite the fact that these are all New World species and starlings were imported and released into the United States only in 1890.[27]

A century has been time enough, it seems, for these immigrants to take advantage of local health care, picking up on medicinal and herbal lore it has taken us millennia to accumulate. Agrimony is famous among the Cherokee as a remedy for ulcers, sore throats and worms. Goldenrod is renowned not just for getting up people's noses, but as a cure for just about everything else. Its generic name comes from the Latin *solidare*, meaning 'to make whole'. And plants of the yarrow genus, named after Achilles, who is reputed to have discovered its medicinal properties, is now a pharmaceutical gold mine – the source of over a hundred biologically active compounds.[56]

There is little that plants such as these have in common as far as visual cues are concerned. There is nothing about leaf shape, size or colour to indicate their chemical affinity. But they are all plants that produce volatile substances which could be useful in fumigating a nest, and should be detectable to the nose of a bird.

Larry Clark at the Monell Chemical Senses Center in Philadelphia has shown that starlings can respond to volatile substances such as butanol, and that cutting the olfactory nerve interferes with their ability to do so.[27] And Timothy Roper at the University of Sussex has found that he can train domestic chicks to distinguish between scents such as vanilla and almond. He suggests that odour memory is important to birds that need to learn how to recognize and avoid toxic foods.[157] Perhaps this is because odour is as potent an aide-mémoire to birds as it is to ourselves?

Mammals are 'supersmellers', with the best noses in the business. Their nasal chambers are honeycombed with sensory areas and extend so far back in the head that they are separated from the brain only by a thin bony plate. And at the front of the face, the nose opens to the air in one of a wonderful array of adaptive structures which range all the way from the flat muscular valve of whales and dolphins to the extended and expressive proboscis of elephants.[185]

The world in which mammals live is laced with all the usual flavours that guide animals to their favourite foods, and mammals are as adept in detecting and tracing these as any fish or reptile. But they have lifted olfaction to a new level of sophistication, giving Jacobson's Organ a whole new meaning and purpose, by sending out as many odours as they receive. Mammals have become scent factories. Their warm blood brews up an aromatic chemistry against which even the most fragrant flowers cannot compete. And these odours seep out into the air through every possible aperture in their soft skins.[4] It is worth cataloguing a few such routes.

The oldest and most abundant odour avenue, which also carries its own external mechanism for effective dispersal, is urine production. This begins, of course, as simple elimination: the disposal of the kidneys' waste products after filtering the blood. But on its way out, urine picks up an amazing range of perfumes provided by the renal tubes, the adrenal glands, the bladder, and the secretions of male accessory sex organs such as the coagulating and preputial glands.

So by the time urine finds its way into the world, it is a very personal product, a mine of information. Male dogs, coyotes, foxes and wolves adopt a typical raised-leg posture that squirts urine high up against vertical surfaces such as trees and rocks, most often when there are rival males around. The enlarged bladders of some species provide a sump for frequent use; for other species it is clear that ordinary urination is distinct from deliberate scent marking.

All tigers of both sexes squat to urinate, letting drops of fluid drip downwards to the ground. But male tigers, like all other felines, have a recurved penis that can also direct a fine spray backwards. And tigers do this far more often, as much as six hundred times more often, than ordinary urination.[145] The product is still urine, but far more aromatic than the usual flow. Its smell is so strong that the Sanskrit name for tiger is *vyagra*, a name derived from a verb root meaning 'to smell'. (This sheds an interesting new light on Pfizer's recent best-selling drug for impotent men, which is being marketed, with or without knowledge of Sanskrit, under the brand name of Viagra.)

The tiger's only rivals in the distribution of urine are the hippopotamus, which atomizes its delivery with rapid fanning movements of its tail; and the black rhinoceros, which expels litres of urine under a pressure high enough to send it five metres or more. This stream is also directed backwards, aimed specifically at a bush which is then torn to pieces by the rhino's horn and spread about, along with a lot of tainted soil, by scraping and shuffling with the feet.

Next to urine, the most obvious and abundant source of odour in mammals is their faeces. These, too, are products of elimination, this time from the gut, but in almost every species they are augmented by secretions from rectal and anal glands. Carnivores are particularly smelly. They specialize in anal pouches, sacs and glands, all of which flavour faeces, but are also often used independently.

European badgers make special latrines which they set up close to the edges of their territories. But they also mark all their usual paths by pressing a gland beneath their tails down in a typical pattern of behaviour which Dutch ethologist Hans Kruuk calls 'bum-pressing'.[105] Brown hyenas leave paste marks on conspicuous grass stalks by backing up to these sharp points and executing what have been described as 'exquisitely delicate movements of the anus'.[138] And zoologist Martyn Gorman describes a behaviour called 'handstanding', which makes it possible for several species of mongoose to rub their anal glands high up on scent posts throughout their range.[65]

At the upper end of the alimentary canal, salivary glands have been pressed into unusual service. European hedgehogs on occasion produce masses of foamy saliva which they flick onto their own spiny backs with the help of a long tongue. This self-anointing happens mainly in the breeding season, but it can be triggered at any time in response to strong smells such as creosote or glue.[15] The behaviour remains mysterious, but like 'anting', in which some birds apply formic acid from live ants to their feather bases, and the manic behaviour of cats in the presence of aromatic herbs like catnip, it probably has to do with sexual attraction. Both sexes do it, and it is said to include secretion from the Organ of Jacobson, though this is the only known example of the organ sending as well as receiving information.

Under certain circumstances, the salivary glands of pigs become so laden with sexual promise that one whiff of a boar's breath is enough to persuade a sow in oestrus to fall immediately into the posture that indicates her readiness to mate.[140]

All these programmed patterns of behaviour in response to the stimulus of a smell are almost certainly mediated by areas of the brain fed, not just by the olfactory epithelium, but also by the busy Organ of Jacobson, in the cause of which no possible new source of smell is ignored.

It would be counter-productive for any animal to court confusion by producing aromatic secretions from its own nose, but several exploit accessory glands of the eye. The only fluid that desert-living Mongolian gerbils can afford to spare, even in spreading sexual scents, is the occasional modified teardrop. Golden hamsters are a little more profligate. With the help of large lachrymal glands, they literally weep hormones.[191] And in some primates, there are vaginal discharges produced by bacterial decay of female secretions that have become so sexually attractive to male monkeys that they have been called *copulins*.[127]

But the largest, most diffuse odour-producing organ in the world is the entire mammalian body: the whole, waterproof but semi-permeable surface of our skins, with all their little fatty and sweaty glands. Glands in the skin are usually associated with hair, and lie inside a hair follicle. But as hair is lost in some species, especially our own, some glands come to lie out in the open, just below the epidermis.

Odours can be produced by any skin gland, but more often come from enlarged glands on special areas of the skin. In roe deer, a patch of these glands appears on the forehead of bucks during the rutting season.[3] In pronghorn antelopes there is a glandular area between the hooves which leaves traces of scent other antelopes can follow, even in the dark, and over stony ground.[131]

In small mammals like shrews, bats, rats and gerbils, skin glands are usually sebaceous, derived from the oily sacs that once conditioned their coats. In larger mammals, there is a mix of sebaceous and sweat glands. And in all sizes of mammals there are pure sweat glands, most often in the angles of the limbs, but also in strange places like behind the ears of white-bellied shrews.[46]

There is, however, one kind of modified sweat gland common to all mammals. Mammals, as their name implies, have mammary glands. They feed their young on milk secreted from these enlarged and specialized sweat glands, and in almost all the four thousand known species of mammals, smell guides newborn infants to their first meal. So all mammals have to have a fully functional sense of smell from the moment of birth.[50] In fact, some mammals even begin to exercise their sense of smell in the womb.

When pregnant rats are injected intravenously with a volatile substance that can be traced, and their foetuses are delivered just an hour later by Caesarean section, they already show signs of the substance in their brains.[141] And the traces in the embryo's brains are found, not in the main olfactory bulb, but in the accessory bulb fed by Jacobson's Organ, which seems already to be part of a system that allows unborn young to get to know the scent of their mother from her amniotic fluid.

The godfather role of Jacobson's Organ continues after birth, helping rabbits, for instance, who otherwise provide their young with remarkably little maternal care. The doe of the European rabbit pulls fur from her own body to make a nest for her pups, but she returns to them only once a day for a nursing visit that lasts no more than four minutes. In her defence, it has to be said that the risk of predators finding the nest burrow increases the longer she stays there. But the naked pups, with their eyes and outer ears sealed, need help if they are going to get weaned and get out before she produces her next litter in less than a month's time.

What young rabbits have on their side of this relentless equation is a fixed-action pattern of behaviour, an instinct that is sparked by an odour and handled by Jacobson's Organ. The trigger is a volatile chemical present in rabbit milk and produced also by the skin around the nipples. Exposed to it when the returning female stands over the nest, the pups begin immediately to shake their muzzles, 'moving across the doe's belly with a sewing-machine-like action until a nipple is reached'. It takes just six seconds for a pup to reach its goal and less than two minutes for it to swallow a third of its own body weight in milk.[80]

This active chemical in a mother's milk is one of the few known examples of what is known as a 'releaser' in mammals: something that triggers a particular pattern of behaviour, and does so every time with the same results.

As far as olfaction is concerned, the big difference between mammals and other vertebrates is that odour has become a very personal business. Fish respond to smells, mammals make them. Solitary reptiles use smell as a way of finding food, social mammals rely on smell to recognize their family and friends. In fact, distinctive mammalian smells play such a prominent role in helping parents to distinguish between their own young and those of other parents that any confusion in this area can be fatal.

Antarctic researchers working with Weddell seals discovered quite by accident that seal pups were rejected by their mothers if weighed in a sack on which another young seal had defecated during its examination. They changed their procedure immediately, and no further rejections took place when they used a new clean sack for every weighing operation.[94]

Newborn domestic lambs run the same risk. They live in large flocks of sheep in which many young are born all at the same time of year, but each newborn lamb is protected by an individual odour with a signature far more complex, than that of baby seals. Part of this is genetic, and part is environmental. The ewes depend on a complex mosaic odour drawn from both sources. A ewe will show interest in a twin removed at birth, or in an alien lamb dressed in a body stocking once worn by their own offspring, but rejects both in the end. And a ewe can be confused about its own lambs once they have been deodorized by scrubbing with a detergent, but will accept them again as soon as the natural scent comes through.[8] All of this presents a host of problems to shepherds and farmers hoping to persuade sheep to foster lambs whose own mothers have died.

The process of recognition by which a parent becomes aware of its young, and vice versa, is called imprinting. It can happen extraordinarily quickly. In domestic goats the first reaction of a mother is to lick the traces of birth fluid and membranes from her kid. She pays particular attention to the flanks and hindquarters, and if the kid is taken away and cleaned before she has the chance to do this herself, she will actively reject it by butting and biting. But if she is allowed just five minutes of normal contact before separation, she forms a lifelong bond with the young animal.[102]

There is a critical period during which such imprinting can take place, and this is far shorter in social animals than it is in solitary species, who generally have more time to get to know each other. In gregarious spiny mice, the window of opportunity for bonding lasts only one hour. If a mother does not have the chance to sniff her pups all over during that hour, they will be alien to her for ever.[146]

Individual animals each have a unique olfactory signature. And as part of this is genetic, so members of the same litter, sharing many of the same genes, come also to have a common olfactory connection. This is very like the 'hive odour' which allows worker bees to recognize one another, or the 'colony odour' of carpenter ants, which originates from the queen but grows into a unique olfactory blend by the addition of genetic and environmental components and allows each worker to recognize another by a split-second sweep of their antennae across each other's backs.[76] Water voles separated from their litter-mates at birth can later be persuaded to accept one another more easily than animals to whom they have no close genetic relationship.[149]

These are not surprising observations. Quite often, smell is the force that prevents inbreeding and encourages populations to keep the sort of distance between them that will eventually translate into the evolution of new species. And in mammals it seems that populations, races, subspecies and species are biological and social categories decided very often, not by genetic incompatibility, but by differences of smell alone.[63]

Discrimination isn't easy: it requires a constant vigilance which, in many mammals, amounts almost to paranoia. Mule deer are distinguished from other deer in North America by their black tails, but they know one another by smell. The odour is secreted from glands on the inside of the knee of the hind leg, and lingers there in a tuft of hairs, each equipped with ridged scales and spaces designed to hold the aroma. Fawns recognize their mothers by sniffing at the knees of several females before making their choice. But all members of a large group of mule deer check each other's leg glands at least once an hour. And if anyone has reason to suspect that there is a stranger in the herd, this knee-sniffing ceremony becomes epidemic, checks being made almost every minute until the outsider is detected and evicted.

Then something even more extraordinary happens. All the deer urinate deliberately down their hocks, soaking the tarsal glands and tufts, and spreading the odour from them by rubbing their hind legs together in a mincing manner, leaving the whole group looking like a party of ladies on high heels in very tight skirts. All this may appear odd, but it has an extraordinary soothing effect on the herd. It accentuates personal smells, but it seems also to meld these into a meaningful group odour, reassuring every deer there that they are among friends.[19]

That sort of cohesion can also be accomplished by deliberately anointing one another. In the forests of Australia and New Guinea, a group of small nocturnal marsupials have taken to gliding from tree to tree in search of the resins and gums on which they feed. Furred membranes that stretch from the forelimb to the hind limb make it possible for these little squirrel-like animals to turn into soaring rectangles, travelling as far as a hundred metres in a single bound. This gliding habit, gently plopping onto the trunks of distant trees in the dark, tends to spread family groups out. So sugar gliders have devised an ingenious olfactory solution.[166]

The dominant male in the group anoints all the members of his colony with a gland on his forehead. He marks females on their chests by hugging them, and they, in their turn, rub some of this secretion off on others, until the entire colony is enveloped in a proprietary aroma that seems to promote general well-being. If a group member strays or fails to stay in touch, it is carefully inspected and energetically re-marked on its return. And if it bears any trace of the smell that identifies a rival group, it is set upon and killed.

The male sugar glider is also an aromatic entrepreneur. Not content with scent-marking his entire colony, he also parlays four further sets of glands into a real-estate venture. With secretions from his throat and feet, he rubs his boundary marks onto strategic rocks and trees, and makes his presence felt even beyond the bounds with random deposits of scented saliva and faeces.

Ring-tailed lemurs manage to do almost as much with a single gland – a thick, hairless patch on the inside of the wrist. With this, the primate which early explorers of Madagascar designated 'the beautiful beast' takes on his rivals. He forearms himself, literally, by rubbing the glands of both wrists over the end of his bushy, banded tail. Then, standing on all fours, he flags this loaded tail above his head, hurling odious threats directly at an opponent in what primatologist Alison Jolly calls 'stink fights'.[90]

The same brachial glands also serve this lemur well in his absence. Ring-tail lemurs live in open country in the south of Madagascar, typically in family groups occupying a fixed territory in a landscape mapped into contiguous domains. Possession of any of these enclaves is a rewarding but exacting business. It depends on the dominant male leaving his mark, a smear from one of his wrists, at every corner of the territory – and on convincing an intruder that he means it. It is a confidence trick, of course, one designed to build his own confidence while at the same time undermining that of a potential opponent. And it succeeds only if he can reinforce the boundary marks at regular intervals – or produce an odour so potent that it seems he has been able to do so.

There is also an element of aromatic politics. Each scent mark carries the signature of a particular individual male lemur, one known to the neighbours on all his boundaries. The neighbours depend, as he does, on a certain amount of diplomatic give-and-take: they are prepared to give him the benefit of the doubt, even if some of his recent markings are not as fresh as they ought to be, and in return he allows them to take small liberties of their own. It is only when the scent of a complete stranger, perhaps a rogue male intent on a hostile takeover, is detected, that all the territory holders go into a higher state of alert, patrolling more often and marking more assiduously.[126]

Scented boundary marks do define territories, but it would be a great mistake to assume that they are simply 'KEEP OFF' signs indicating private property. More often than not they are far more complex communications, tending to inform and educate in the best tradition of good advertising. Many animals scent-mark their territories, but there is little evidence to show that such olfactory fences ever prevent intruders from entering. What they do instead is to provide familiarity, boosting the morale of a resident who, as a result, enjoys a 'home ground' advantage in any conflict there with a stranger.

Jean Jacques Rousseau, whose individuality bordered almost on megalomania, believed that the builder of the first fence was the founder of civilization. Two centuries later, we know more about animal behaviour and can see how many species depend on declaring, marking and defending a territory in some way. Most such areas are clearly visible. You can even see some of them from the air. But the fences most mammals build are invisible, because they are entirely olfactory.

They may also be enormously complex, not only demarcating an area, but redefining it altogether. Domesticated dogs, for instance, are famous for urinating on lamp-posts. But they don't urinate on *every* lamp-post. They may sniff at each one, with results ranging from growling, through ground scraping, to defecation. But counter-marking with an overlay of new urine remains the response of choice.

This behaviour is generally assumed to serve one of two functions: either the top scent obliterates all others, giving a territorial advantage to the latest performer, or the scents meld into a sort of composite group odour, identifying the presence of a society. But recent laboratory studies by Robert Johnson at Cornell University suggest a third and more interesting possibility.[89]

Golden hamsters of both sexes scent-mark by arching their backs and rubbing their sides, where there are well-developed skin glands, against vertical surfaces. They prefer, as dogs do, to deposit their scent where others have done so before them. And tests of familiarity with smell signatures show that, while hamsters pay immediate attention to the top scent, they are well aware of previous markings and can even identify these as individual odours.

So hamsters perceive the marking post in a very complex fashion. At one level it is a bulletin board, a place to pick up the latest information about everyone in the area. But at a deeper level, they also learn from the frequency, volume and quality of all markings something about the local power structure. The post is not just a display board, but a ladder board, giving some idea of who is likely to win the current tournament.

A common feature of all mammalian chemical signalling is its transparency. It is not a secret system, but one open at least to all members of the same species. Often the news spreads to other species, and not necessarily through the sense of smell.

Most markers are placed as high as possible. Bears advertise their territories by scraping bark off trees and rubbing against the bare surface, reaching up to their full and clearly awesome height to do so. The high calcium content of the droppings of the spotted hyena turns them bright white in the sun, and they can be seen from a distance. The object seems to be not just to advertise occupancy, but to convey information about the occupants.

Far from being repelled by such marks, most intruders of all species are strongly attracted to them, going out of their way to take a sniff. In most of the natural world, no news is *not* good news. Any news, even bad news, is better than none. And all news, good or bad, gives the bearer a distinct selective advantage.

Only mammals make actual scent marks. Reptiles may leave an odour trail behind them, but before mammals arrived on the scene no animal actively put down chemical markers on chosen places, or made such extensive social use of simple chemical signals.

Mammals have turned chemical sensitivity into a major sense system. The olfactory tissue in their noses has spread out from its confines on the roof of the upper chamber and now covers a larger part of the nasal cavity, encroaching on the non-sensory area of scrolled turbinal bones where incoming air is warmed and moistened. An antelope of our body weight may have sensory epithelia ten times the area of ours, with at least ten times as many sensitive cells. Amongst carnivores, the disparity may be even greater. Only cetaceans, who have chosen to return to the ocean, smell more poorly than we do.

There is no doubt that the rest of the mammals, small or large, have a sense of smell more acute, more active, more subtle than our own. But these differences are largely ones of degree, a function simply of their larger, more efficient noses. But there is another way in which mammals have changed the rules and the practice of olfaction in a qualitative manner, which might not put us at quite such a disadvantage.

Reptiles favoured Jacobson's Organ and smelling through the mouth. Mammals built on the reptilian model and, beginning with the mammal-like reptiles, brought the nasal system of smell up to at least the same standard. Our olfactory cells and the olfactory bulbs have flourished, continuing to feed and inform the cortex with news of the world at large. But because mammals are more actively social, in closer contact with one another, they have also invested heavily in a new, improved way of detecting and responding to smells that carry information of a personal nature.

Mammals have the best developed Organs of Jacobson in the world, and exploit these, very largely, for their social potential, for acquiring information about other members of their species that cannot be gathered in any other way. And humans are still very much part of this global 'Odornet'.

Ambrosiacos

Ambrosia lies at the heart of Linnaean thinking about scents and smells.

It has its roots in amber, something resinous, fragrant and attractive, known to the Greeks as elektron. When rubbed, it releases charged particles, which became regarded both as the elixir of life and as nourishment for the gods. Sappho sees it as a balmy drink, Homer as the food of the immortals. Milton's paradise is anointed by a deity whose 'dewy locks distilled ambrosia'. And everyone agrees that it is divine.

Linnaeus is more down to earth. His paradise has Comus, the god of sensual pleasure, distributing ambrosial oils 'through the porch and inlet of every sense'. And he chooses to describe this group of odours as sweet and heavy fragrances, redolent of musk. He sprinkles this term liberally about, setting up a little Oriental deer as Moschus; and condemning a garden of aromatic plants – germanders, geraniums, sages, mints and mallows – to be for ever musky or moschatum.

But in the end, no bloom can compete with mushkas itself, with the golf-ball-sized pod which the Asian deer carries under the skin of its stomach. For more than five thousand years, this has been a manifestation of the gods, incorporated alike into perfume bases and the foundation of cathedrals, so that both bodies might breathe out its bewitching odour.

It is still worth its weight in gold.

3

On the net

Hitler started it all. In the early 1930s the German biochemist Adolph Butenandt, then still under thirty years old, began the work on human hormones that won him a Nobel prize in 1939. He was forbidden by Hitler to accept that award, and turned instead to another line of research.

It was already known that moths could be sexed by the shape of their antennae. Those of the females are generally simple, but male antennae tend to be elaborately feathered, decorated with twenty thousand or more minute hairs, each designed as a receptor for a particular chemical message in the air. This signal originates in a gland on top of the female's abdomen that produces a volatile substance which the male seems to find irresistible, even from many kilometres away.

Butenandt set out to isolate this 'essence of attractivity', and turned to the silkworm moth for answers. That was a fortunate choice, because silkworms were already being farmed on a large scale in Italy and Japan, and it took half a million female moths to produce enough of the secretion for him to analyse. It also took twenty years of patient extraction and fractionation before he isolated from it a substance capable of exciting a male moth at a distance. But in 1959, Butenandt was finally able to announce the discovery of *bombykol*,[21] a substance so powerful that if any one female moth were to release all of her store in a single spray, there would be enough to bring a trillion males to her side.

In the same year, endocrinologists Peter Karlson and Martin Luscher decided that the time had come to distinguish between hormones – organic substances that travel within the body to produce their effects – and chemicals such as bombykol that are released into the environment and influence other individuals of the same species at a distance.[93] From the Greek verb roots *pherein*, 'to transfer', and *hormon*, 'to excite,' they coined the influential new name of *pheromone* for something capable of exciting at a distance.

Pheromones are definitely exciting. Diane Ackerman calls them 'the pack animals of desire'.[1] Once released, they enjoy an existence quite separate from the organisms that produce them. Volatile ones travel freely through the air, delivering their messages, having profound effects at a distance. Stable ones settle on rocks and trees, lingering there long enough to exert their influence at another time. They all work a wonderful organic magic, in mammals very often through the medium of Jacobson's Organ.

In no mammal is the opening to Jacobson's Organ obvious. It is not like a flared nostril, but tends to be tucked away somewhere on the edge of the airstream. The olfactory sense area in the nose is equally isolated, and needs to be stimulated by active sniffing, pushing the head forward and taking a quick short breath that carries air deep into the recesses of the nasal cavity. But Jacobson's Organ has boosters of its own.

Most obvious of these is an unmistakable facial expression: a curl of the upper lip, the sort of grimace we make when faced with an unpleasant odour. It is common in deer and horses, especially among males exposed to the urine of a female in oestrus, and has come to be known as the *flehmen*, or 'flared' face. The behaviour is accompanied by a thrown-back head which is so conspicuous that it has become part of a visual display, but its true purpose is pheromonal.[49] By contracting the lip-raising muscles, pressure is brought to bear on the nasopalatine canal, which runs from nose to mouth, opening it and exposing the ducts of Jacobson's Organ.[73] Close study of male goats in flehmen, which wag their heads from side to side, suggests that this behaviour also allows fluids in the mouth to flow across the lower opening of the canal that leads to the organ.[109]

Flehmen is most common in ungulates, but it crops up in some surprising places. Bats are anomalous animals. They have become so well adapted to the dark that hearing is now their major source of information. And like birds, their ability to fly puts them beyond the reach of most useful environmental odours. So bats have converted their noses almost entirely to sonar systems and now have very little sense of smell, and little sign of Jacobson's Organ – with one important exception.

Common vampire bats feed entirely on mammalian blood, which they acquire by tiptoeing up on their sleeping prey. Their muzzles are unusually large for a bat, partly to house a set of razor-sharp teeth which slice into the skin of horses or humans so delicately that the cut is seldom felt. But the stealthy approach is critical. Young bats must learn the necessary skill, and frequently follow adults onto the same prey, which the adults mark with a fine spray of urine.[165] As they approach, all the bats flare their upper lips, sniffing their way to the source, picking up the rich smell of blood and urine, letting the pheromones play over their Jacobson's Organs – which are by far the largest in the order.[32]

Another surprising flehmen face features the largest nose in any land animal. Elephants may be tall, standing well above the critical odour zone, but their elongated trunks put their nostrils back down there on the ground, where scents abound. Bull Indian elephants respond to the scent of female urine by dipping the tip of their trunk into the liquid and, with a lip curl quite unlike any used in normal eating and drinking, apply this directly to the roof of the mouth. There, in a strange return to the old reptilian pattern, Jacobson's Organ opens directly as two clearly visible pits, about two centimetres apart.[154]

Elephants and vampire bats both have well-developed muscular tongues. Along with most kinds of cattle, they are able to use their tongues to compress the hard palate and force fluid from the mouth through the nasopalatine canal directly into Jacobson's Organ. If you look just behind the upper front teeth on most mammalian skulls, you can see the bony channel that houses this canal – large in cows and tigers, vanishingly small in seals and ourselves.

Most other mammals rely instead on an internal, automatic version of such a 'pump'. In the noses of golden hamsters, cats and guinea pigs, there are several large, thin-walled blood vessels lining the cavity of Jacobson's Organ. When these constrict suddenly, a slight vacuum is created, sucking nasal fluid into the organ itself. When they fill with blood and expand, fluids are expelled, setting up a tiny two-way 'vomeronasal pump' which ensures that the organ receives the stimulation it needs.[125]

It seems clear that in most mammals the machinery already exists for receiving pheromonal signals. This is fortunate, because 'you've got mail'. The air is full of messages, and we are all born to be subscribers to the Odornet.

Sex in mammals is aromatic, intimate, and only deceptively simple. Boars, both wild and domestic, have musky breath. Steroids in their saliva combine to produce a pheromone with a predictable effect. Any receptive sow that gets a whiff of it stands motionless with her ears cocked, her back down and her rump in the air. This posture is called lordosis, from the Greek for 'bent back', and it can be induced even by an agricultural aerosol marketed as 'Boar Mate'. Used correctly, it works, showing that even in mammals, behaviour can be released by a chemical signal.

But sometimes signals get crossed. For centuries, truffle-hunters in the broadleaf forests of France have used pigs to help them find the precious fungus, which is attached to oak roots as much as a metre underground. Traditionally, a muzzled sow was pressed into such service, perhaps because she was easier to handle than a boar. In 1982, German researchers discovered that the musky smell of a ripe truffle is caused by steroids: not just any steroids, but a combination identical to the one that boars produce and store in their salivary glands. The best and most-prized black truffles, found in Perigord, prove to contain twice as much pig pheromone as the blood plasma of even the most lusty boar.[121] That is what attracts the sows, which immediately drop into the lordosis posture, pointing like bird dogs at the spot where the odour is most apparent. Very nice for the truffle-hunters, but what's in it for the truffle?

It may be that the answer is to be found in another agricultural problem. Half a century ago, farmers in Western Australia became alarmed by a sudden decline in the fertility of their sheep. Between 1941 and 1944, the numbers of lambs born each year fell by 70 per cent. There was a world war in progress, and little could be done then to find out why, but suspicion fell on a new crop of subterranean clover, and the problem came to be called 'clover disease'. Since then, similar outbreaks of infertility in sheep, cattle and rabbits have been linked with common pasture plants, mostly members of the legume family, such as clover and lucerne. And in the last twenty years the chemicals responsible have been identified. They are all related to plant pigments called flavenoids.[171]

In mammals, a hormone produced by the ovary is responsible for the growth of the female reproductive tract, and of the mammary glands, and for the coordination of mating behaviour. It is called oestradiol, the primary oestrogen, and recent research reveals that its biochemical structure is almost identical to that of the flavenoids in clover. How did this come to be?

There is only one thing that clover and sheep have in common. They are partners in a plant–herbivore relationship: one eats the other, and the other does what it can to avoid being eaten too often. The clover produces an oestrogen, a chemical that mimics a key mammalian hormone, and this acts as a contraceptive pill, reducing grazing by reducing the number of grazers.

That a plant can do this is not just amazing, it is rather disturbing. It means that the capacity to produce even the most complex chemicals is universal. If clover can create an oestrogen to which sheep respond, and truffles turn out pheromones designed to attract the attention of passing pigs, there would seem to be no limits. The traditional barriers between animals and plants, or predators and prey, all fall away in a world where even the most potent elixirs are passed around like canapés at a cocktail party. And the fact that some of these key chemicals are volatile puts everything, literally, up in the air.

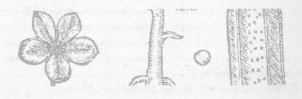

Pheromones are everywhere. They are woven into the atmosphere like invisible threads in a great aerial tapestry, setting up the sort of universal chemistry that goes a long way to support the old idea of a common sensory system.

Most known pheromones consist of just eight or ten carbon atoms linked up with other elements in a limited number of ways. But the permutations possible within this array are large enough to produce very precise instructions. They tell us, in the words of Lewis Thomas:

When and where to cluster in crowds, when to disperse, how to behave to the opposite sex, how to ascertain what *is* the opposite sex, how to organize members of a society in the proper ranking orders of dominance, how to mark out exact boundaries of real estate and how to establish that one is, beyond argument, one's self.[192]

The very simplicity of pheromones is both their strongest and their weakest point. They can be produced by almost any living thing, but they can also cause confusion. And now comes news that pigs are not the only ones being duped. Michael Kirk-Smith and his colleagues at the University of Buckingham have discovered that the truffle steroid is not just mimicking pig pheromones. It is a match also for a steroid we produce and secrete through the sweat glands in male armpits.

Truffles are described as smelling like 'musky nuts', which is not exactly the flavour of the week at ice cream stalls. But this odour may be nothing more than a marker, something olfactory to jog the memory and remind us of how good we felt last time we smelled it. In Birmingham, UK, volunteers exposed to the truffle steroid consistently gave photographs of the opposite sex far higher scores for attractiveness than did subjects who breathed nothing but filtered air.[98]

Now you know why truffles are expensive!

Pheromones are indeed pack animals, all-purpose chemicals that can quickly be adapted to many needs. They are extremely useful for connecting individuals, often over great distances, but they do not necessarily offer an exclusive and secure system of communication. The lines are open, but they are also open to abuse.

In Colombia, female bolas spiders hang from trees in the forest, lying in wait after dark, armed with a sticky ball on the end of a single line of thread. They send out a volatile chemical signal in the form of a pheromone which precisely imitates the sexual attractant of the female army worm moth. And any male moth that answers the call runs the risk of being snagged and reeled in. The only solution to such piracy is privacy. The initial attractant can be generic, broadcast at large and designed specifically for the nose. But the next step needs to be different, more intimate.

To attract a male, the female golden hamster lays a scent trail with a secretion from her vagina that smells a little like broccoli, but contains more than two hundred volatile components. Alan Singer at the Monell Chemical Senses Center in Philadelphia has isolated one of these, a well-known smelly substance called dimethyl disulphide, and found that just a few molecules of it are sufficient to attract a passing male hamster, who uses his nose to home in on the broadcasting female like a moth. Then the game changes.[174]

The male and female hamster meet head to head, and something about him, perhaps a cheek gland odour, catapults her into lordosis. Then he moves around to nuzzle and lick her rear end, picking up a new smell from a protein that seems to be aphrodisiac. This is *not* volatile and is channelled directly to, and analysed by, his Jacobson's Organ. With it, he mounts and copulates. Without it, or deprived of the use of an organ to detect it, he loses interest altogether.

This is where the alternative system comes into its own, dealing with odour sources too large or too stable to become airborne. It makes whole protein molecules accessible to smell, and brings an entirely new range of substances into the business of carrying social signals – substances that can only be transferred by direct contact between two individuals. Imagine what havoc would be caused by an aphrodisiac that was simply turned loose in the air, triggering mounting behaviour in every male hamster in the region. This way, mating takes place only between the intending pair.

There are three basic ways in which a smell can influence reproductive behaviour:

The first is as an *initiator*, where a pheromone sends a signal that tells an individual something about the gender and breeding condition of another. Boar's breath does that at close quarters; a female cat or dog in heat does it at a distance.

The second is as a *primer*, where an odour produced by an individual starts a sequence of long-term physiological change in others. The odour of a male hamster, for example, can be seen to shorten or synchronize the oestrus cycle of all females in the area.

The third effect smell can have on reproduction is as a *terminator*. This is a variation of the primer effect and occurs, for example, when pregnancy in mice is blocked by exposure to a strange male.

As it happens, all three effects are mediated by, and involve, Jacobson's Organ.

Most of the experimental work on Jacobson's Organ in mammals has been done on small rodents. It began in 1975 with a short but influential study of hamsters by Bradley Powers and Sarah Winans at the University of Michigan.[150] They compared mating behaviour in male golden hamsters deprived of smell, or of Jacobson's Organ, or both. Hamsters with no ordinary sense of smell mated normally, while those without their Jacobson's Organ were severely impaired, and those with neither olfactory sense failed to mate at all. In another, more recent study, bank voles deprived of their Jacobson's Organs could not even tell the difference between males and females.[104]

Most rodents demonstrate the 'Coolidge effect', named for the thirtieth President of the United States, who entered biological folklore as a result of a visit to a government farm in 1928. He and his wife were taken on separate tours, and when the farm's champion rooster was introduced to Mrs Coolidge with the comment that the bird mated several times a day, she said, 'You must be sure to tell that to the President.' When the President himself was told of the rooster's prowess, he wanted to know first if the bird performed always with the same hen. And when informed that the rooster was presented with a new hen every time, the President nodded thoughtfully and responded, 'You must be sure to tell that to Mrs Coolidge.' Sexually satiated male hamsters can always be revived by a novel female, but not if they have been deprived of their Jacobson's Organ.

Smell is everything in the sex life of pigs. The attraction between a boar and an oestrous sow is purely chemical and is not diminished by taking one animal out of the other's sight, or even by anaesthetizing it. But without olfactory bulbs in the brain, neither pig can distinguish between a boar and a sow on the basis of sight or sound alone.[172]

Beyond mere sexual identification, the most important information provided by reproductive pheromones is the sexual condition of the sender. The vaginal secretions of female monkeys become obviously attractive to males just before ovulation, but there are measurable changes in the relative concentration of different fatty acids throughout her cycle. And it is likely that all males in a troop are very aware of, and affected by, this progression.[127]

They may even be partly responsible for it. In several rodents, the mere presence of an adult male induces oestrus in females.[194] The trigger is a pheromone in his urine which is volatile and easily dispersed. But South American acouchis, long-legged wild relatives of the guinea pig, make sure of their influence by actively spraying urine over the female during courtship. There is no evidence that this encourages or excites her in any way. On the contrary, on the occasions when I have seen this behaviour, the female seemed to find it rather disconcerting. But as she licks the urine off, it is picked up by her Jacobson's Organ and appears to kick-start rapid oestral development. It also sends a very clear signal to rival males. It acts on them like an aromatic beacon announcing that this female is 'owned' and has, in all probability, already mated.

In addition, a large number of male animals make absolutely certain that their presence is smelt by spreading urine over their own bodies. Goats deliberately urinate onto their stomachs. Reindeer spray their hind legs. Camels anoint their tails, with which they spread the urine all over their flanks. And many antelope urinate on the ground and then roll in it. The action of these priming chemicals is species-specific. The urine of another kind of goat, deer, camel or antelope won't do.

In addition to simply inducing oestrus, there are three other effects of pheromones on reproduction, each named after its discoverers:

The *Lee–Boot effect* comes from Dutch researchers who showed that when female mice are kept entirely away from males, their oestrus cycle starts to lengthen. This suggests that the normal hormonal cycle depends in some way on the presence of a male pheromonal coordinator.[195]

The *Whitten effect*, from Jackson Laboratory in Maine, extends this with the finding that in an all-female environment, menstrual cycles collapse altogether as mice experience a kind of pseudopregnancy, which can be reversed only by the reintroduction of a male.[208]

The *Vandenbergh effect* from North Carolina State University rounds out this trilogy of male hormonal influence by demonstrating that the mere presence of any adult male in a mouse colony speeds up the sexual development of pre-pubertal females.[194]

But it must be borne in mind that these are all laboratory effects. It is highly unlikely that any rodent population anywhere in the world would be totally deprived of male company. All that these results show is that pheromones in the urine of male rodents actively participate in coordinating many aspects of their reproduction.

Larger mammals get their news about the breeding condition of individual animals from different sources. Male Indian elephants have a temporal gland just behind the eye which, in season, produces a secretion known as 'musth'. This happens to males over twenty years of age, and when it does, a bull uses his trunk tip to smear the product over himself and his surroundings. Cow elephants find it irresistible and use their trunk tips to transfer the odour directly to the openings to Jacobson's Organ in their mouths. And as the effects of the musth pheromone spread through a herd, they coordinate oestrus in the adult females.

The muskier odours of some mammals are obvious even to our noses and must contain an individual ingredient, because those of a stranger block pregnancy, interfering with implantation and growth, sometimes even inducing an abortion. In prairie voles, the introduction of an alien male to a colony results in the termination of 80 per cent of early pregnancies there. Later in gestation, the incidence of terminations drops to just over 30 per cent, but litter sizes still suffer.[180] There seems to be nothing directly terminal in the odour of any male. It is his 'strangeness' alone that produces these effects, but the stranger can and does mate with the resident females soon after their loss – which, presumably, is the whole idea.

Generally speaking, these odour effects work as well as they do because females have a lower threshold for the catalytic smells than males: they can detect them, nasally or through Jacobson's Organ, at far lower concentrations. In humans, musk goes totally undetected by half of the world's men – usually the ones who wear too much cologne – while women are able to detect musk even at dilutions as low as one part in a billion. And women's sensitivity to male hormones reaches its peak precisely at the moment of their ovulation. During menstruation everything changes, and some women become as anosmic as most men.

Sigmund Freud was quick to dismiss our sense of smell in either sex, attributing our lack of acuity to the distance of our noses from the ground. There is something in that, but Freud was, of course, thinking and writing before odourless pheromones were discovered.

Very few actual pheromones have been isolated and analysed in any mammal. In this respect, we know more about silkworm moths than we do about ourselves. Much of the time, we use the word 'pheromone' as a concept, a sort of biochemical mantra, something to bridge the gap between ignorance and understanding. There is nothing wrong with this, as long as we pause from time to time to remind ourselves that 'pheromone' may not mean a single active agent. It could equally well describe a combination of chemicals, or a mix of chemistry and behaviour.

For example, an olfactory cue may be directly responsible for sexual arousal. This seems to happen with pigs, a boar's breath sparking a specific response in a sow. But such fixed-action patterns can be dangerous, even lethal, if there is no mechanism for taking account of other circumstances. A sow that freezes in the lordosis posture could be very vulnerable if she is surrounded by a group of rival boars, or under observation by a tiger.

The pheromonal message travels from her Jacobson's Organ to the accessory bulbs in the forebrain, and then directly to the limbic system where sexual behaviour is triggered. But fortunately, most mammals also possess a rival smell system. Messages picked up by the nose travel to the olfactory bulbs, and then on to the cortex, before finally ending up in the limbic area. This indirect route is vital, for it allows the sort of modification of response that can make a difference to a sow in potential trouble. It can make all the difference in the world for males of most mammalian species, who are very prone to behave irrationally in the presence of attractive females. The neural delays at least provide an opportunity in both sexes for second, and perhaps more reasonable, thoughts.

So what we mean by 'pheromonal activity' in such cases is a very subtle and integrated pattern of reception and assessment in which both olfactory systems have a role. But there is still more to the process. Cues that are described as 'primer pheromones' also involve ductless glands such as the pituitary, which produces the hormones that regulate metabolic functions and bring about a whole range of reproductive behaviours. And all this elaborate biochemistry comes under the scrutiny of those parts of the brain which are responsible for the imprinting of useful olfactory information, the sort of information that allows a mother to recognize the smell of her own offspring, or all females to recall the odour of their resident male.

In the end, the most direct way of demonstrating the real importance of Jacobson's Organ is to deprive an animal temporarily of its use – or to remove it altogether.

Charles Wysocki and his colleagues at the Monell Chemical Sense Center have pioneered a procedure for removing Jacobson's Organ from small mammals.[215] The Philadelphia technique involves use of a dental burr to drill through the hard palate directly into the organ's sleeve in the septum, and scraping it out entirely. Care has to be taken to remove every last cell on both sides of the septum, because these regenerate themselves, and many individuals seem to be able to restore sensitivity with just 3 per cent of the original tissue. Without an Organ of Jacobson:

Male house mice show no increase in testosterone after meeting a female, and produce none of the ultrasonic calls which normally reassure a mate.

Male guinea pigs fail to make the characteristic head-bobbing courtship gestures that usually follow exposure to female urine.

And male prairie voles stop urine-marking their territories, and are far less aggressive towards rival males.

Female house mice no longer have the ability to delay puberty in their nest mates.

Female prairie voles fail to produce lordosis and therefore to copulate.

And female guinea pigs, already mated, fail to recognize their offspring, and make no effort to retrieve them when they wander from the nest.

Olfactory cues from the nose are still available to all those animals. But it is abundantly clear that, without Jacobson's Organ, which provides a privileged and direct pathway to those areas of the brain concerned with sexual behaviour, these cues alone are insufficient for normal reproduction.[215]

The primary role of the nose seems to be analysing smells with no predetermined meaning. The Organ of Jacobson specializes in recognizing smells that carry specific information about gender, reproduction, and dominance status. And the fact that Jacobson's Organ is situated out of the mainstream, hidden in the septum of the nose, could have something to do with avoiding inadvertent stimulus and inappropriate behaviour. It looks, sometimes, as if we mammals may need protecting from our own organs.

It is also becoming clear that, though the two mammalian systems of olfaction may be parallel in structure, they are often complementary in function. They work together, producing a synesthetic effect within the single sense of smell. This may help to explain the extraordinary sensitivity of dogs.

With the exception of breeds whose faces are so flat that they have chronic respiratory problems, all domestic dogs smell well. Their world is a multi-layered extravaganza of odours, very different from our own, and the ones who know it best have long, drawn-out snouts filled with scrolls that multiply our postage-stamp size olfactory sense by a factor of perhaps fifty.

Sheepdogs have 200 million olfactory sense cells. German shepherds 250 million, labradors 280 million and beagles over 300 million receptors, all with the fringe of microscopic hairs that increase the active surface area of each cell several times. Any of these animals possess a sense of smell thousands or maybe even millions of times more acute than ours. But the masters of odour are pedigree bloodhounds, whose feats put them in a league of their own.

Bloodhounds are not exactly good-looking. They droop. Their skins are several sizes too large, hanging loosely about their necks and heads like dewlaps. Even their eyelids sag under the weight of eyes that look so ravaged they could belong to an opium addict. Their ears are long and leathery, with an eighty-centimetre span, designed not so much to hear as to sweep up scent as the hound moves its head from side to side. And the mouth is almost invisible, masked by jowls and curtained by a continuous drool.

All this slobber and saggy flesh are there for a single purpose. Centuries of breeding have selected for these features because they work: they absorb odours near the ground and collect and cup any scent that lingers in the air. There is a record of a hound in Bennington, Vermont, which succeeded in picking up a trail eight days old, following it into a grocery store and a bank, sniffing the ground and high up on bushes and buildings, crossing busy streets until it ended at a bench outside the local bus terminus. The missing man confirmed later that he had sat there briefly before boarding a bus to California.[31]

Bloodhounds find us naked apes easy prey. Every day we shed forty million flakes of skin, each one supporting a rich flora of bacteria with characteristic odours. All this invisible and odorous dandruff drifts along behind us in a vapour trail, leaving a track that leaps out at a hound like neon lights. Even on a quiet day, we produce several pints of sweat, the smell of which is so distinctive that most dogs can retrieve a pebble from a river bed and return it to the hand of the man who threw it. In a laboratory test, a dog showed that it could identify the single glass slide in a box that had been touched briefly by one human fingertip six weeks previously.

The reason that dogs can manage such feats may have a lot to do with the fact that their Jacobson's Organ is large and uniquely equipped with sensory cells that are not bristled, but carry hundreds of cilia, very much like those in the nose. There is nothing like this in the Jacobson's Organ of any other group of animals. The same strange cells may exist in wolves, jackals, coyotes and other wild canids, but their organs have yet to be examined under scanning electron microscopes.[2]

It looks as though those doleful dogs, the bloodhounds, who would as soon lick a quarry to death as bite him, have put together the perfect odour detection system. It links a large nose full of sensory cells, all designed for picking up volatile traces, to a Jacobson's Organ equipped with customized tissue tailored perhaps to the reception of larger particles. And the combined effect of the two olfactory systems acting in tandem allows everything to make good scents to a hound.

The total area of a bloodhound's olfactory tissue amounts to something about the size of this page. But even this expanse can become saturated. Smell cells get used to an odour and are quick to tune it out, which is why a smell that is overpowering when you enter a room can soon become almost unnoticeable. We adapt to it. Familiar smells are no longer novel, and that can make them dangerous.

A bloodhound that becomes bored with a smell would be useless. It needs to follow one chosen scent until it finds its prey. To do this, it must inhibit the natural tendency to shut off a familiar smell after a decent interval. It has to focus on that single stimulus beyond all the usual limits of sensory attention spans.

And to accomplish this, it is possible that bloodhounds have evolved Jacobson's Organs which are good enough at the usual functions of the nose to alternate with it, giving each system the chance to take a breather, to stop smelling for a while, without interrupting the chase for even a moment.

The Beagle Brigade, whose dogs remember the smell of, and sniff out, prohibited imports at international airports in the United States, take no chances with their charges. They guarantee a 90 per cent success rate in detecting drugs by giving their sniffing dogs twenty minutes off in each working hour.[130] But humans are different. One of our best tricks is to learn how to suppress, rather than encourage, the conscious recognition of an odour. We encourage the suppression of smell awareness in order to pay attention to other things. We do something that most species cannot do: we turn sensitivity to certain smells off when it suits us. And this ability has had the direct result of increasing our intelligence in other ways and at the expense of our sense of smell. We may now even lack the genes necessary to restore our old olfactory acuity, but the equipment we still have is more than sufficient to keep us on the Odornet, where smell continues to play an astonishing and unexpectedly potent role in our lives.

PART TWO

THE FRAGRANT APE

'Smells,' said Rudyard Kipling, 'are surer than sights or sounds to make your heart-strings crack.' Just so. Things you see and hear fade fast, but where smell is concerned there seems to be only long-term memory. All of us, no matter who we may be, have the same olfactory apparatus and the same kinds of sense cells. But we may well remember the sensation in very different ways.

People in other cultures don't necessarily share our traditional construct of there being five senses.[28] The Hausa of Nigeria divide the senses into just two categories, with sight in one and everything else lumped together in the other. And in some Buddhist systems, mind is classified as the sixth sense. Even in the West, there is no real agreement about the number. Plato added senses of heat, cold, pleasure, discomfort, desire and fear to the list. Philo of Alexandria felt it was necessary to include a religious sense, one he called 'the Love of God'. And in early medieval Europe, speech was seen as a vital sense, a gift from God.

There has never been any agreement about the ranking of the senses. Aristotle put sight at the top, and followed it with each of the other senses according to the position of their organ on the human body. Diogenes the Cynic, who went around in daylight with a lantern, looking for an honest man, gave precedence to forthright smell. And Pliny created a whole race with nothing but a nose on their faces, able to live on scent alone.

Arguments continued until the Enlightenment, when animal

senses in general were discounted in favour of the intellect, and Descartes could bring himself into being simply by thinking. But it was the English philosopher John Locke who dispensed entirely with spiritual subtleties and laid the foundation for modern scientific studies of perception, by insisting that it was through sensory experience that ideas reached the mind. And somewhere between sense and sensation, the potent influence of culture lies waiting.

In the Andaman Islands, time is conceived of as a succession of odours. The seasons of the year are named after the fragrant flowers in bloom at the time – a calendar of scents. Among the pastoral Dassanetch of Ethiopia there is also a succession of odours, alternating the sweet smell of new grass in the rainy season, with the acrid smoke of burning old growth in the dry; the good and bad smells of life and death, renewal and decay. For both peoples, odour marks the passage of time and defines their living space, their home, as part of a familiar smellscape.

In tropical rainforest, where vision is limited, smell becomes even more definitive, setting up olfactory zones. There is little or no wind under the canopy, and odour lingers so long that very often you can smell farther than you can see. There is a musty background scent of rotting vegetation, but anything over and above that, the smell of a campfire or of a wild mammal, becomes very conspicuous and provides useful information.

In the Colombian Amazon, the Desana are so aware of such odours that they even call themselves *Wira*, which means 'The People Who Smell'. And they do. The Desana are hunters and seem to acquire the musky smell of the game animals they eat. Their neighbours, the Tapuya, are fishermen and carry ready evidence of their trade. And the nearby Tukano, who are farmers, are said to smell of roots and freshly turned soil.

The Suya Indians of the Brazilian Mato Grosso take such olfactory distinctions even further. They recognize differences even within their own community, and say that adult men smell 'bland', old men and women 'pungent', and all young women 'strong'. This classification is not unlike that provided by Linnaeus, who grouped his seven types of odour into three similar categories: ones that smell pleasant, ones that are positively unpleasant and ones in the middle that can go either way. But in Suya society, the distinction is a purely social one which uses odour to rank its members.

In a sense, all systems for classifying odours are arbitrary. There is such heavy social, cultural and historical baggage attached to terms like 'pleasant' and 'strong' that they have no real meaning outside the context in which they are being used. And when it comes to personal smells or body odours, no one outside your own group is ever going to smell as good as you do. It is a truism amongst researchers into smell that all human subjects behave as if they themselves do not smell like humans, because all humans smell bad. There is never going to be any global agreement about this, either.

European languages are particularly weak in their vocabulary of smell. Although we can recognize thousands, perhaps even hundreds of thousands of smells, all our descriptions of them are based on a very limited range of terms borrowed from taste. And in English, we even use the same word, 'smell', for an odour, *and* for the process of perceiving that odour, *and* for the unfortunate circumstance of emitting such an odour.

Non-European languages are not necessarily any better. The Borro in Brazil fail even to distinguish between smell and taste. But their neighbours who speak Quechua, the old Incan tongue, exercise a nice discrimination in confining their word for smell to the act of actually inhaling an odour. There are seven other words for describing the acts of smelling something good, smelling something bad, smelling something edible, smelling something 'fishy', smelling something in company with other people, letting oneself be smelled and being forced to smell someone else.

There is similar sensitivity in a widespread belief that a link exists between odour and personal identity, something that might be called a 'scents of self'. In the Malay Peninsula, the gentle Temiar people equate personal odour with the life-force and take great care not to impinge on another's odour envelope. Disturbing the 'odour soul' is thought to cause illness. So anyone who passes too close to someone else or has, for any reason, to invade that individual's smell-space, recites 'Odour, Odour, Odour' as a combined warning and blessing.

If there is any one thing that is true about human response to odour everywhere, it is that we are ambivalent about it. Biologist Michael Stoddart calls this the Zoological Conundrum:

I believe that the most curious feature of our sense of smell is that while we generally relish the sweet scents of a summer garden, or the bouquet from a fine wine, we do not generally relish the natural scents of our fellows.[187]

This is true. We are well equipped with odour-producing glands, but find the odours they produce embarrassing. Humans, it seems, must not smell like humans. So we go to extraordinary lengths to eliminate or disguise our own humanity. The idea that people should smell of people is generally regarded as subversive; anyone who smells too obviously of themselves is probably a danger to society.

This distaste percolates through our languages, turning almost any references to another's odour into a form of verbal abuse. Those of whom we disapprove are 'rotters', 'stinkers', 'foul-mouthed' and generally offensive. Their only hope of reconciliation lies in total olfactory denial and a rush to smell of something else. And the ultimate irony is that, in this flight from ourselves, we cloak our own reproductive identity in the secondary sexual odours of other species. We are clearly at odds with our own sense of smell – and need to know why.

Tetros

Linnaeus was well aware of our olfactory confusion. Midway between desire and repulsion, he created a category of fragrances described as 'tetrous', from the Latin word meaning that which can be offensive to the senses, or possibly even foul. 'Full of gross humours' was the euphemism of the time.

There is a strong flavour of social disapproval in all discussion of what odours belong here, making for some strange associations. So we find tomatoes and potatoes grouped together with opium poppies and belladonna, love apples and moon flowers in the same bed. Appropriate, perhaps, for a classification whose translation ranges from 'polluted' through 'shameful' to 'obscene'.

Tetri may smell foul, but they sell well. The Devil always did have some of the best tangs.

But there is something grand, too, about the grouping, which gives pride of place to the doughty walnut: the tree and the seed Linnaeus dubbed Juglans, from the glans of Jove or Jupiter's Nut. There is no doubting the symbolism of its shape or the fact that the putrid smell of the kernel is produced by a chemical called juglone that is also antiseptic, herbicidal, and a very useful treatment for some tumours. You can't judge anything by its odour alone.

4

Scents of self

We may be the naked apes, but we have more skin glands than any hairy ape – or any other mammal, for that matter.

Our two square metres of skin, enough to make a bedspread, comes equipped with three million sweat glands that can secrete over eight litres of fluid a day. Most of this is coolant evaporating to keep the surface of the body from overheating, but there must be more to it than that. Mixed in with the saline solution of sweat are numerous amino acids.

Sweat glands are simple, tubular structures. They are most numerous, four times more dense than anywhere else, on the palms of our hands and the soles of our feet – right where they need to be, if we are going to leave our mark on the world around us. The next greatest density is on the forehead. This is a strange place to perspire, unless cooling is less important than conveying something significant to those with whom we meet face to face. It would have to be important to make it worth the risk of having salty sweat run into our eyes. Or could bushy eyebrows be nature's version of the sweatband?

No one has yet isolated an active pheromone in human sweat, but a study in 1977 involving the clean hands of pairs of people found that subjects can distinguish very easily between male and female hand odours, even though hands possess only sweat glands. Telling two men or two women apart was harder, but some subjects were even able to separate identical twins, as long as they had been on different diets for at least three days.[201]

The odours of garlic and asparagus are quick to appear on the surface of the skin, but there may also be more emotional compounds in sweat. Enough, perhaps, given the chance to incubate in confined spaces like shoes or gloves, to give real pheromonal satisfaction to foot and hand fetishists. And because we all sweat more under the influence of fear or passion, when our whole metabolism speeds up, sweat is certainly the vehicle for signals produced by other kinds of glands.

Apart from sweat producers, we have two additional skin glands on areas other than the hands or feet. The first are the sebaceous glands. They produce a thick, clear, oily secretion, rich in free fatty acids, which provide most of our distinctive odour signature. Some sebaceous glands are independent, opening directly onto the surface of the skin on lips, eyelids and nipples. Most, however, are tied to hair follicles, even if the hairs themselves have almost disappeared. The original function of the glands was to condition and waterproof the fur, protecting it from becoming soggy. And they continue to work that way on the scalp, but most sebaceous glands are now dedicated to the production of social and sexual odours.

The largest ones can be found in association with our finest, silky hairs. On the forehead, alongside the nose, in the midline of the chest and around the anus, where hair is hardly visible, our most productive sebaceous glands release their products. And the fact that most of them do no work at all until we reach puberty suggests that the messages they carry are predominantly sexual. Those in men are always bigger, and a direct result of the ducts of these glands becoming blocked in either sex is an outbreak of acne. Generally speaking, in those areas where they do occur, sebaceous glands are not grouped into special olfactory organs.

Apocrine glands are so grouped. These coiled structures, which are large enough to be seen with the naked eye, also empty into hair follicles. They gather densely on the scalp, around the pubic area, and in the navel, but their greatest concentration is in human armpits.

There is little in the animal kingdom, with the possible exception of the musk deer's pod or the civet's anal gland, to compare with the human armpit for olfactory potency. It is our major source of body odour, an organ perfectly designed for the job. Apocrine glands there are heaped up, two or three to a follicle, covering a patch the size of a tennis ball, coating the long underarm hairs with their oils, vaporizing easily in the warmth, dissolving and spreading with the help of sweat glands that keep the whole area moist and bacterially active.

The secretion of apocrine glands is different from the sebum of sebaceous glands. It is a viscid oil, coloured anything from a milky white to a blood red, depending on diet and perhaps on race. Japanese of the nineteenth century, when first exposed to European traders, described them as *bata-kusai* – 'stinks of butter'. People of European and African ancestry do have the largest armpit or axillary organs, often so densely packed with glands that they look like sponges under the skin. People of Asian origin have smaller organs or no armpit glands at all. In Japan, 90 per cent of the population has no detectable underarm odour, and young men who are unfortunate enough to belong to the smelly minority can even be disqualified from military service on that ground alone.

Apocrine gland secretions, on their own, have no smell. The musky odours of the armpit are all entirely the work of bacterial decomposition, which turns fatty oils and hormones into phero- mones. Men produce a stronger odour than women with the same bacterial flora on their skin. And women, as their hormonal levels change during oestrus, produce dramatic changes in their own armpit odours. This is all to do with shifts in basic chemistry, and it is surprising that deodorant manufacturers haven't yet begun to market a range of products for every woman. Calendar Cosmetics: 'One for every day of the month. You're worth it!' But they will, as they already have with a range of products for different national markets, taking local preferences into account. In Central Africa alone, axillary odours are described as 'phosphoric, cheesy, nutty, garlicky, rancid, ammoniacal and musky' – and not all of these qualities are thought to be undesirable.

The fact that such descriptions often end up sounding more like tastes than smells is further evidence of the paucity of smell words in all our vocabularies. But there is one descriptor that applies to underarm odours almost everywhere: there is general agreement that they can be very sexy. The French novelist Joris- Karl Huysmans puts it well in this description of one of his heroines: 'The scent of her underarms easily uncaged the animal in men.'[82]

It is no accident that armpits should be the odour glands of choice in human encounters. They are closer to other human noses than the rest of our apocrine centres, and can be turned on or off at will by simply raising or lowering your arm. Almost every gesture we make spreads the sorts of odours which, depending on your circumstances, are either attractive or repellent. It is significant that in many welcoming and most submissive gestures both arms are held up to give a stranger or an opponent the best chance of gauging your real intentions: 'If you don't believe me, smell me!'

Stories abound, too, of rustic swains who work all day in the fields with a kerchief under one arm, and wreak havoc after dark with this message-laden cloth in a breast pocket, carefully placed right under a dancing partner's nose. In rural Austria it is apparently still the practice for girls to keep a slice of apple in their armpits during dances, and to present it at the end of the evening for the pleasure of the man of their choice. And word has it that in 1572, Henry III of France inadvertently wiped his face on a sweat-stained chemise, abandoned between dances in the cloakroom at the Louvre by the beautiful young Mary of Cleves, 'and from that moment, conceived for her the most violent passion'.[187]

Mary, of course, would not have shaved under her arms. She wasn't part of a culture like ours that supports research to show that '24 hours after thorough cleansing with Ivory soap, only one out of ten shaved axillae could be described as odorous, compared with nine out of ten unshaved armpits'.[168] In their day, even princesses took advantage of the fact that underarm hair provides a large surface area on which bacteria can breed, and acts also as an efficient wick, encouraging the broadcasting of that odour every time the arms are raised, as they must be when dancing with a taller man, or when two lovers put their arms around each other and she snuggles into his shoulder.

Our culture requires that we frown on underarm odour and fight rearguard battles against it with deodorants. So there are no paeans to the armpit in poetry and song, but no shortage of sentimental requests to put heads on, cry on, or lean on, obliging shoulders. Once we were not so squeamish. 'I will be arriving in Paris tomorrow evening,' Napoleon wrote to Josephine. 'Don't wash.' Which might translate today as: 'Don't disturb the wonderful process of incubation that takes fresh apocrine secretions and allows aerobic diphtheria and other micrococcal bacteria to process it into a priming pheromone with a chemical configuration not unlike musk.'

Napoleon said it better. Perhaps he had secrets of his own. We know he drenched himself with eau de Cologne before battle. But that wouldn't have prevented non-odorous steroid precursors in his armpits from metabolizing, with the help of time and his own bacteria, into odorous steroids which are almost identical to the primer pheromones on boar's breath. Small wonder, then, that male armpits in France were once known as 'spice boxes', or that whiffs of them can still exert an attraction, both strong and animal, that seems to be beyond our cerebral control – partly because it goes straight from the axillary organ to Jacobson's Organ.

Apocrine glands everywhere on the human skin are the main source of what we now call body odour: the dreaded 'BO' that upsets people in crowded trains or locker rooms. But that same bacterial stench, in different circumstances, carries vital information. When volunteers in England agreed to wear T-shirts for 24 hours without bathing or using deodorants, every one of them was later able to identify their own shirts by smell alone, even when put with two others. Most were able also to identify the sex of the volunteers who had worn the other samples.[159]

In the United States, Richard Doty at the University of Pennsylvania followed this up with a series of carefully controlled studies. In these, underarm odours were collected from men and women who wore gauze pads in their armpits for 18 hours. The pads were then presented in 'sniff bottles' to male and female judges who were asked to rate them for strength and pleasantness. They agreed on which were strongest, and correctly identified these odours as male, but failed to agree on whether they were pleasant or unpleasant.[43]

This is where the social, sexual and cultural ambivalence begins.

At University College in London, students were asked to evaluate applicants for a college job while wearing surgical masks lightly impregnated with a male hormone, androstenol. Others wore clean masks, and all were told only that the purpose of the mask was to prevent the judges from being influenced by one another's facial expressions. The results were interesting. Female evaluators exposed to a trace of musky smelling androgen gave higher assessment marks to male applicants.[34] And when Tom Clark of Guy's Hospital in London followed this up by spraying androstenol inside several telephone booths at a mainline railway station, women spent more time there on the phone than they had when the booths were left unscented.[44]

At Birmingham University in the UK, a related male hormone called androstenone was sprayed on a seat in a dentist's waiting room, and the pattern of patient movements was recorded on the following days. This androgen has no perceptible odour, but in a practice with equal numbers of patients of either sex, far more women used the seat than men, and all seemed more willing than other female patients to wait their turn.[99]

It looks as though these hormones, whether or not they can be consciously detected, do provide us with information about other individuals, and that such awareness is not peculiar to any one culture. In 1981, a large cross-cultural study was made with German, Italian and Japanese subjects. All wore cotton shirts for seven consecutive nights and used no perfume or deodorants. All were then given ten shirts and asked to identify, by smell alone, their own shirt and that worn by their sexual partner. They were also asked to sex the remaining eight shirts and to grade all ten for pleasantness.

Most had no difficulty in picking out their own and their partner's odours. And many were able to sex the remaining odours quite accurately, with the women in all three cultures being more accurate in this regard than their men. All the cultures agreed that male smells were stronger and less pleasant than female smells. But Japanese women differed markedly from their European counterparts in finding even the smells of their own male partners unpleasant.[164] This is a clear reflection of taboo in a culture where few people have underarm glands and most bathe frequently. But on the whole there is a remarkable agreement across all these national and cultural boundaries on the gender of odours, and the fact that friends smell better than strangers.

There is also evidence that armpit odours can go a long way towards making friends of strangers. At Hatfield Polytechnic College in England, first-year students were enrolled in a project which required them to wear a special necklace overnight. Some of the necklaces were impregnated lightly with musky androstenol, and all the volunteers returned the next day and described their social interactions during the time spent wearing the test accessory. Those who wore control necklaces with inert substances in them noticed no change in their circumstances. But women students, after a night of exposure to the active ingredient of male underarm secretions, found that they were far more responsive to men the following morning. They were generally more willing to make contact with strange men, and more successful in doing it.[35]

It looks as though androstenol can induce an 'approach' response in human females, and should on this evidence be regarded as a possible releasing pheromone in the reproductive behaviour of our species. This doesn't put it on the same level as boar's breath, or suggest that our behaviour is as easily manipulated as that of pigs. But it does imply that underarm secretions can indeed influence the behaviour of others.

And it becomes necessary to look with renewed interest at things like a Victorian device sometimes called the 'love seat'. This was a small sofa built in an S-shape which allowed a young couple to sit close enough to talk to each other, but facing in opposite directions, and in no possible danger of physical contact. The true reason for its popularity becomes apparent only when you realize that this was a very shrewd pheromonal piece of furniture, putting a suitor's armpit within inches of his intended's nostrils, without in any way offending contemporary proprieties.

It is no surprise that women are better than men at detecting and recognizing personal smells. They need to be. Richard Porter and Jennifer Cernoch at Vanderbilt University in Tennessee have been exploring human odour signatures for years, beginning with mother–infant relationships. They put identical shirts on twenty newborn babies for 24 hours, then took the garments from the children and presented them singly to the mothers in cardboard containers with a small opening at the top. Each mother's task was to guess, by smell alone, which container held the shirt worn by her infant. Sixteen of the twenty mothers did so immediately, with 'absolute certainty', regardless of race, age, the sex of the baby or previous experience of parenting.

This experiment involved breast-fed babies who, after normal deliveries, had up to six days of contact with their mothers. A second test was confined to babies delivered by Caesarean section who were bottle-fed and had spent an average of only two hours in direct contact with the mothers. The results were only marginally less impressive, thirteen of the twenty mothers making the right choice, first time.[147]

There are only two ways such recognition could take place. One is by a genetic similarity of odours, in which the mother recognizes her own smell in that of the child. The other is by early imprinting on the child's individual odour, in the same way that rabbits and sheep learn to bond with their offspring. Both would seem to rely on direct, rapid, and unthinking contact with the mother's limbic system – which suggests the involvement of Jacobson's Organ.

I suspect that both processes play a part. In the first few days of a human infant's life, when its cries are only intermittent and it may be invisible in the dark of night or in the gloom of a simple dwelling, only smell can give it a unique identity. And only the baby's birth mother has the opportunity, and the genetic advantage, necessary to make a quick connection. She already knows half the olfactory story. She may, of course, also get a little help. The mother–infant relationship isn't a one-way street.

Kittens recognize the odour of their mother's milk, and before long the smell of a single teat. Two-day-old human babies, even while sleeping, make a sucking sound when exposed to the smell of their mother's breast. If a breast pad worn by the mother is placed in its crib, the baby turns towards it very quickly, ignoring a clean pad, or one that has been worn by another lactating woman. The pheromonal system appears to begin operating at a very early age, just as it does in rabbits, and it spreads quickly from breast odours to those of other parts of the body.

Jennifer Cernoch and Richard Porter tested two-week-old infants, presenting them with a choice between armpit pads worn by their mothers and by another woman. The babies showed clear interest in their own mother's smell, preferring it to the smell of any other woman, lactating or not, but only if they were already being breast-fed. Bottle-fed babies not only miss the breast experience, but also spend far less time exposed to their mother's bare skin, and most show very little interest in any of her odours.[25]

That deficit can have far-reaching consequences. Any direct response by an infant, including preferential smiling and reduced crying in a mother's company, produces a reciprocal and equally positive response from her. And what starts with simple odour recognition builds quickly into a complex social relationship fuelled at every level by smell.

Whatever else may take place between mothers and their children, they will always smell alike. In Nashville, a test was run with family pairs consisting of mothers and their five-year-old children. Each individual was given a standard T-shirt and asked to wear it for three nights. On the fourth day all the shirts were collected and sealed in containers, then offered one by one to test subjects whose task was to match each target shirt with its partner in a family pair. The choice lay between the garment worn by the right mother or child, and three others. The chance of being right is one in four, or 25 per cent. But the average score in a long series of tests was over 50 per cent, showing that even strangers can identify family relationships by smell alone.[148]

Families smell familiar, which gives us all a standard against which to match the odour of others. They don't smell like we do. And the fact that our similarity to one another is genetic rather than environmental is nicely shown by a version of the Nashville T-shirt test, which involved pairs of husbands and their wives. Despite the similarity of their diet and lifestyle, no one could pair the garments which belong to such an otherwise unrelated couple.

Our body chemistry is largely genetically determined. In this respect, we are no different from house mice or honey bees, which makes it appropriate to leave the last word on domestic odours to e.e. cummings:

> all good kumrads you can tell
> by their altruistic smell.[38]

Anthropology in the last century was obsessed with studies of kinship, with meticulous accounts of the complex relationships between tribal people connected by marriage and descent. Genetics in this century has simplified the problem, turning blood relationships into a branch of mathematics. A child shares half its genes with a sibling or a parent, a quarter with a cousin or a grandparent, and so on. But some of the elaborate kinship systems established by tribal people were so sophisticated that they not only anticipated certain modern DNA tests, but went on to expand such awareness of relationship to include all the kinds of affinity that can occur between people who might not be of the same blood, but nevertheless need to live and work together. And they could do this, I believe, long before Gregor Mendel worked his mathematical magic with pea plants, and made us aware of heredity, because they were already secure in their knowledge of who was directly related to whom. They could smell the difference.

Most of us still can. A number of studies have shown that mothers can discriminate among their own offspring by smell. All children share half of their genes with each parent, but they are not genetically identical. Each is a genetic and olfactory individual, except, of course, for identical twins, who come from the same fertilized egg. A study in 1955, however, showed that dogs are able to tell even identical twins apart by their smell, as long as the odours of both are given to them simultaneously for the sake of fine comparison.[92] They, and I suspect we, are capable of doing what genetics cannot. We have a very subtle sense which makes it possible to cleave to a long-lost second cousin, even when everything else about them may be unfamiliar. The smell is the thing.

Our new knowledge of genetic advantage, and the way in which behaviour can be modified by relationship, shows that most of any calculation of affinity is made unconsciously. We know some things, and do some things, because they feel right, not because we know that a blood relationship exists. And we avoid doing some things for precisely the same reason.

One of the most obvious selective advantages of kin recognition is an avoidance of inbreeding. In our species, as in many others, there seems to be a taboo against incest. The closer the genetic relationship, the stronger the taboo becomes. Male house mice exercise such a constraint on the basis of odour alone, preferring to mate with unrelated females, whose smell tells of a certain essential genetic distance. And it is possible that we too make use of olfactory clues in such circumstances.

Closely related men and women in the same household are subject to the same pheromonal signals as strangers. But something stops them from responding in the way a stranger can. The exact nature of the moral block remains mysterious, but studies in Israel on kibbutz children, who grow up in groups of their peers, offer a useful hint. One survey concludes that 'there are no marriages between people who have been continuously reared together for their first six years', whether or not they are related by blood.[169]

Living together when you are young binds you to one another in a special non-sexual way. Living together later doesn't have the same result. There is no evidence that college students in co-educational halls of residence, or husbands and wives in their own homes, are any less likely to get sexually involved simply because they become very familiar with each other. The 'kibbutz effect' suggests instead that incest taboos are maturation phenomena. They kick in at puberty, because we seem to have a gene which programs us, from that time on, to ignore the sexual attractions of those with whom we have grown up, simply because these are the ones likely to be most closely related to us. And because these attractions are at least partly to do with odour, it is likely that the block is also an olfactory one.

Sexual behaviour is governed by the limbic system, so any modifier of it is likely to be there, too. Something very simple which just says, 'Stop! Ignore that smell. It comes with another very familiar smell that means the message is not for you.' In such circumstances, it is best for everyone that the Organ of Jacobson is the only organ involved.

Inspired by the Vandenbergh effect, in which male mice have a hormonal effect on females, George Preti and Winifred Cutler of the Monell Chemical Sense Center in Philadelphia exposed women with irregular menstrual cycles to the smell of male underarm sweat.[151] All the women had cycles less than 24 days or more than 32 days in length. But after having an extract of male hormone rubbed under their noses several times a week, all the women's cycles moved close to the 28-day norm. And away from male influence, such regularity disappears in humans, just as it does in rodents or elephants.

There is evidence also of male pheromonal influence on the age at which human female menstruation begins. In our species, menarche can start anywhere between nine and eighteen years of age, depending on both genetic and environmental factors. Poor nutrition, some illnesses and even intense athletic activity can all delay the onset of puberty. But in the last two centuries a new factor has advanced the age of first menstruation by three months in every decade, or almost three years per century.[20] In other mammals, puberty in females is delayed by exposure to adult females, and accelerated by exposure to adult males. And the same seems to be true of us. Ever since the Industrial Revolution, the pheromonal climate of our homes has changed as more women have gone out to work, and more men have spent more time in the homes of smaller families.

No matter when menarche begins, all human females are spontaneous ovulators, and it cannot be coincidental that the usual cycle in our species is almost identical to the period between two full moons. Such 'lunacy' is not uncommon in biological rhythms, but it does not, of course, mean that all the world's women menstruate on the same day. But they could.

The necessary research began in the late 1960s, when Martha McClintock was an undergraduate at Wellesley College in Massachusetts. She noticed that women in her dormitory often menstruated on the same days, and wondered if there was survival value in such coordination. So she began working with undergraduates at the all-female college, asking two simple questions: 'When did you last menstruate?' and 'Who are your two best friends?' The results surprised everyone, were widely reported, and soon became known as the 'McClintock effect'. 'Women who spend a lot of time together,' she concluded, 'tend to menstruate at the same time.' But no one knew why.[118]

Almost a decade passed before anyone else had the courage to touch such a contentious subject. But in 1980 the same crucial questions were posed on another college campus, this time a co-educational one in Britain. The answer was the same: best friends are more likely to menstruate and ovulate on the same days.[66]

The next link in the chain of evidence came that same year,

from California. Michael Russell and his colleagues at the Sonoma State Hospital picked up on the possibility that one of the women in a group sharing a menstrual date could be a 'driver', synchronizing the others. They rubbed an extract from the underarm of a woman who had a very regular 28-day cycle under the noses of sixteen others, three times a week for four months. And at the end of that period, all of the test group were menstruating within three days of the donor, who acted just like a male coordinator in the Lee–Boot effect discovered in mice.[160]

It seems that odours from one person can influence the menstrual cycle of others. And there is good reason to suppose that hormonal activity in humans can be brought under olfactory control. But getting anything that smacks of folklore and old-wifery accepted as science is not easy. What the McClintock effect needed was a convincing biological advantage, something to explain why synchronized cycles might be useful and have survival value to humans. The benefit, as far as rodents are concerned, is that pregnancies in any colony can be achieved by a single dominant male, and births brought together to coincide with seasonal peaks in the availability of food. In humans the benefits are less obvious, particularly in the culture in which McClintock made her initial discovery.

In the West today, more women work away from home in male company, and more men spend more time in the home. Family sizes are smaller, and there are fewer places where women spend time in the exclusive company of other women. Polygamy, of course, still exists elsewhere, perhaps in over two hundred societies, in most of which the wives continue to live together in separate quarters.[61] But in both East and West, in developed and developing nations, one thing remains constant.

Adolescent girls gather together in groups in which there is a great deal of close contact and tête-à-tête, Jacobson-to-Jacobson activity, creating perfect conditions for pheromonal communication. And because all of them are around the age of menarche, and experiencing early menstrual events which tend to be erratic, irregular and often fail to lead to proper ovulation, they could benefit enormously from any influence that is likely to coordinate, regulate and discipline their unruly physiologies.[97] Add to this the recent discovery of a male cycle in which men show a regular rise and fall in body temperature and the production of essential steroids, and we have most of what we need to make the McClintock effect universal and useful.[48]

Male bonding activities, such as sports, now begin to look like essential physiological coordinators instead of just rowdy guy stuff. Pheromones are being exchanged like crazy in all the scrums, rucks and tackles, on and off the field, adding hormonal maturity and psychological well-being to the bruises. And now that we know that the production of testosterone and other androgens is often synchronized with the menstrual cycles of men's wives and lovers, the whole Jacobsonian process acquires the metronome it needs, especially among those of us who live in the concrete canyons of modern cities, to put our biological clocks back in time with the rest of the planet.[142]

During the last decade, some depth has been added to the study of pheromonal communication among humans. Aron and Leonard Weller of Bar-Ilan University in Israel have shown that what they call the common dyad of mother-and-daughter-living-together is a potent one. It results in a significant degree of menstrual synchrony, and demonstrates that such cohesion is even easier between those who already have history and chemistry in common.[205] They have also brought this research into the workplace, looking for evidence of synchrony among women soldiers thrown together in the Israeli Army – and finding it there, too, but only among women who are also close friends.[206]

It is not surprising to find that pheromonal contact is improved by close social interactions. The Organ of Jacobson is, after all, a rather personal device. But the real breakthrough, bringing at last a measure of scientific acceptance to Martha McClintock after almost thirty years of official scepticism, came with a paper published in *Nature* in 1998.[182]

It is a lean, elegant publication, answering all the critical questions raised by McClintock's early work. This meticulous study involved twenty women who had pads wiped under their noses every day for four months. Some of the pads were blank, some carried irrelevant odours and some were impregnated with odourless underarm secretions from other women. But the important feature of these tests lies in the timing of the chosen human secretions. Those taken from women who were menstruating, *shortened* the previously regular menstrual cycles of the subjects, by as much as 14 days, while those taken from women who were ovulating, *lengthened* the menstrual cycles of the subjects by as much as 12 days.

There are obviously two complementary chemical signals involved here. One accelerates ovulation, the other delays it. And between the two, oestrus cycles balance out at the natural, lunar period of about 28 days, showing that volatile underarm secretions from apocrine glands of one person can, and do, regulate the biological rhythms of another. This helps explain, too, how a group of teenage girls, all with irregular menstrual patterns, can arrive at an arrangement which brings them into line. The two complementary pheromones, working in opposition, make the essential magic of meeting in the middle, along a line which synchronizes all of them with one of Earth's most fundamental rhythms. All the girls need is enough of them to push the system towards equilibrium – a menstrual quorum.

So there are strong arguments for the existence of human pheromones. And all that remains to be done, to finally convince the doubters, is to prove that the odourless substances are perceived and passed on by olfactory tissue or by Jacobson's Organ.

I have been concentrating on the legendary armpit, but it is hardly the only source of human odour. We are prone also to having 'bad breath'. Richard Doty and his colleagues at the Smell and Taste Research Center in Philadelphia started with the assumption that – unlike armpit odours, which tend to be strongest in men because they have more apocrine glands and seldom shave their armpits – there ought to be no anatomical advantage to either sex when it comes to breath odours. Their chosen subjects were all dental students with good oral health, who ate the same food and used no toothpastes or mouthwashes on test days. The judges sat behind a screen and sampled breath blown through a glass tube into a sampling funnel. They were asked to guess the sex of the breather and to rate the breath's unpleasantness.[44]

Most judges were able to get the sex right, with women again scoring better at the task than men. And all agreed that the odour of male donors was more intense and less pleasant. No one actually mentioned 'boar's breath', but male saliva does contain androstenone, which is in itself odourless, but is more easily detectable by women and more likely to be repugnant to men. It is also known that the sense of smell in women becomes more acute around the time of ovulation, when oestrogen levels are at their highest, and that sensitivity then is particularly high for sex-linked pheromones.[64] All this makes sense and suggests that, after all, a sigh is *not* just a sigh, and there have to be suspicions, too, about a kiss.

In evolutionary terms, mouth-to-mouth contact begins as infant care, with bird-like passing of food, sometimes chewed or partly digested, from parent to young. African wild dogs of both sexes regularly regurgitate meat for all the pups in the pack. But in many species the gesture has become a natural part of courtship in which the members of a pair investigate each other's mouths. The kiss of chimpanzees is somewhat chaste, but their cousins the bonobos have made it patently erotic. 'Bonobos, like human beings, walk arm-in-arm, kiss each other's hands and feet, and embrace with long, deep, tongue-intruding French kisses.'[55]

For them, as for us, such behaviour is most common during bonding, at those times when every aspect of intimacy needs to be explored before real commitment can occur. A kiss is a way of bringing into play Jacobson's Organ and the limbic system as well as the more thoughtful cortex. The kiss takes over where no sigh can go. It is an excellent, self-rewarding, private and palatable way of transferring the sort of chemical signals that are too heavy to be volatile, but still play a vital role in coordinating reproductive behaviour. So, happily, the fundamental things do still apply.

Kissing on the mouth remains a very personal exchange and, in all other circumstances, it is significantly modified or deflected onto cheek kisses or 'air' kisses, where olfactory contact is minimized and Jacobson's Organ is excluded. And wherever the kiss has been adopted as part of human submissive behaviour, the lower the rank of the kisser, the lower his kiss has to be. Catholic bishops kiss their Pope on his knee, lesser mortals have to be content with kissing an embroidered cross on the Papal shoe, and heathens and the penitents in many traditions are required to go all the way and 'kiss the dirt'. It is, in the end, simply a matter of getting pheromones to their proper places.

If the popular magazines are to be believed, women attach more importance to kissing than men do. With their greater sensitivity to smells, and their comparative lack of salivary hormones, it is likely that women do, in fact, get more out of a kiss than men can. But the tables are turned when other parts of the body are involved.

In every primate ever observed in the field, males are very obviously attracted to the genital area of females at certain stages of the oestral cycle. Male macaques, mangabeys and baboons all investigate females at regular intervals by touching their vagina with a finger and sniffing it. As ovulation approaches, marmosets, woolly monkeys and all the apes join them in licking and smelling vaginas directly. And at ovulation, when females are both more receptive and most sensitive to odours themselves, everyone becomes immersed in mutual genital investigation.[187] Humans are prone to the same behaviour.

At the Monell Chemical Senses Center in Philadelphia, human vaginal secretions were sampled every day, collected on sterile tampons and sealed in glass jars. Then those from every stage of each donor's cycle were presented at random to male and female testers, who were asked to evaluate them for strength and pleasantness. The results were conclusive: all testers found secretions collected close to menstruation most strongly scented and least pleasant, and all agreed that secretions produced during ovulation were distinctly different.[42]

Human vaginal secretions are highly complex, containing more than thirty compounds, mostly fatty acids; some odour-bearing, some not; but all subject to bacterial decay. Taken together, and in a series of combinations that vary dramatically throughout the cycle, there is sufficient information in such secretions to enable anyone with an active nose and sufficient curiosity to predict the time of ovulation and behave accordingly. Monkeys certainly do. The jury is still out on humans, but studies show that intercourse in our species takes place more often near ovulation than at any other time, whether or not the man is conscious of his partner's cycle.[45]

They also show that where such activity is initiated by the woman, it is almost always during the hormonal surge that accompanies ovulation. It would be surprising if this were not so: the genetic goal of copulation, after all, is procreation. And it would be equally surprising if men did not respond positively to an odour that coincides with both ovulation and sexual opportunity, whether the smell is pleasant in itself or not.

Where sex is concerned, we can be led by our noses. But it is unlikely that only one pheromone is responsible. The complexity of the menstrual cycle, and the number of substances involved, suggests that any search for the ultimate aphrodisiac, the one great sex attractant for our species, a human bombykol, may be fruitless.

Sweet Nellie Fowler, a celebrated courtesan in Victorian England, made a fortune selling handkerchiefs which she had taken to bed with her, though history fails to record where exactly about her person she tucked them.[178] The mystery was undoubtedly part of her allure. And mystery remains part of our continuing curiosity about each other's odours, and helps to explain the popularity of oral sex in an era when most other body parts have been deodorized and depilated out of olfactory existence.

Athletes in training, or before a big game, are often advised to avoid sex. There is little evidence, however, to suggest that intercourse has any direct influence on athletic ability, except perhaps by depriving the participants of sleep. But there are a few professions with which sex can create havoc. Wine tasters, tea blenders, and perfume testers are all familiar with a condition known as 'honeymoon rhinitis', which leaves a sufferer with such nasal congestion that smelling, and therefore tasting, becomes almost impossible. Opera singers are leery of it for the same reason, and perhaps not without good cause.

The temperature of the nasal mucosa has been found to rise by 1.5 °C immediately after sexual intercourse, as a direct result of rapid dilation of the veins there. No one knows why, but this is one of a syndrome of sympathetic nervous responses that accompany orgasm.[51] Sex goes, it seems, not to the head, but to the nose. Some people get nose bleeds, others sneeze and, after a temporary hypersensitivity to smells, many feel congested. The Roman physician Celsus recommended that, at the first sign of a cold or catarrh, men should 'abstain from warmth and women'.

The naso-genital relationship seems always to have been on our minds. Virgil tells of adulterers being punished by amputation of the nose – a fitting punishment, perhaps, given the importance of smell in the sort of sexual attraction that begins in puberty with an explosive growth of hormones. In this sense, smell is a secondary sexual character, and the nose one of our most important sex organs, and one that may even have had to be constrained at a critical stage in our evolution.

Michael Stoddard at the University of Tasmania is one of the very few to have considered the role of olfaction in our early history.[186] He points out that our distant ancestors lived in small family groups, as many primates still do today. But sometime in the Miocene, starting perhaps ten million years ago, major changes took place. We became more vertical, more bipedal, more mobile, more predatory and less hairy, and began to live in larger groups.

As a result, males took on more responsibility for feeding the family, and joined with other males in pursuit of the best high-calorie resources. Females, faced with the painful problem of giving birth to babies with increasingly large heads, shortened their pregnancies and found themselves having to deal instead with longer periods of dependence. Both these changes put a greater premium on keeping pair bonds together and, to that end, sex began to play a larger part in our lives, over and beyond that of simple reproduction. So the sexes came to look very different from each other. Males grew bigger in every way, until they had the best-developed genitalia in any primate. Females became sexually receptive for longer periods and advertised their maturity by developing breasts in advance of lactation. And both sexes became more odorous than any other primate has ever been.

Sex was very much in the air. It was in everyone's noses and uppermost in their minds. Lives depended on it, but there was a problem. To make it possible for teams of men to go out on hunting forays, the group grew to more than fifty individuals. Some of these were adult men who needed, or chose to, stay with the women and children. And that put a new and unexpected pressure on the pair bond.

Humans are the only mammals who bond in pairs and live in large groups. It isn't natural, and Michael Stoddard suggests that evolutionary pressures dealt with the problem by finding a way of allowing private sex to continue at a high level, without attracting too much attention from rival males. And that the solution that arose was to be more secretive about ovulation, and to tinker with our software in ways that changed the way we feel about body odours.[186]

There is something in this idea. It would help to explain our continuing ambivalence about our own smells, our insistence that humans should not smell like humans at all, and our tendency to seek out and borrow the smells of other species. We men certainly have become so alienated from the natural cyclic odours of our women, that we have become uneasy about them, saying that we find them repugnant and making them the focus of some of our most inflexible taboos.

The truth, however, is that we continue to be very smelly apes. Our armpits and vaginas go on producing strong, musky odours, and we carry with us all the olfactory apparatus necessary to perceive them. If there is a switch that turned the tide, and turned part of our sensitivity off back there in the Miocene, it was clearly a mental one: something in the newer parts of the big brain.

The old, wide range of olfactory cues continues to wash over us, but it is obvious that we no longer respond to them in quite the same way. We have, in a sense, adopted a reptilian solution to the problem, downgrading the sensitivity of the nose. We have, in effect, gone into a state of partial olfactory denial. But, like the reptiles, we still have access to a second system, one that has been allowed to persist because of its more private nature. We have a small, but apparently functional, Organ of Jacobson that operates at close quarters with a different, more down-to-earth array of messages. And it is becoming very subversive.

Most of the action stimulated by Jacobson's Organ takes place at an unconscious level. Its effects are felt primarily in the emotional centres of the brain, spreading outwards through the pituitary gland and the gonads. It has the capacity to bypass the neural blockage and produce the sort of feedback that brings things once suppressed back to conscious attention. And it has the sneaky habit of reminding us that the old strictures, the mind games of the Miocene, are no longer absolutely necessary. With it, perhaps we still have the chance to regroup, to re-enchant our lives and become more aware of the sexy world of scent out there, just waiting to be rediscovered.

Nauseosos

This is it – the ultimate stench, a name for anything disgusting, loathsome, offensive and likely to put you off your food.

Linnaeus drew attention to the Latin root nausea, originally applied just to seasickness, but soon spreading to anything of ill savour. He was an early supporter of homoeopathy, noticing that the odious and highly toxic hellebore, when taken in minute quantities, was a remedy for pain and convulsion, the original 'hair of the dog'.

The ancients defined all things nauseous as having 'animal humours' and recommended them as a treatment for 'excitements'. And the excitable poet laureate John Dryden lost all his positions and pensions for frequently and unwisely describing those in power in England as nauseating.

The nature of things truly nauseating ranges from 'the stench of an old sepulchre' to 'stale salt fish', but the most convenient source was a plant growing on the sandy plains of Afghanistan which produced a foul-smelling resinous gum, known locally as asa foetida.

But the botanical epitome of nauseosi are the South African stapelias. Their carrion stink attracts blowflies which become so entranced with flowers looking like mould growing on rotting meat, that they lay eggs on them. One species compounds the gut-wrenching illusion by adding minute hairs that give the entire plant the appearance of seething with tiny flies.

You wouldn't dream of eating a stapelia. And that might be the whole point . . .

5

In bad odour

The biggest killer in the world is not war or famine or cancer. It always has been, and still is, an ancient scourge known as 'bad air' or *malaria*. Sewage disposal and water purification have gone some way toward lessening our paranoia about airborne pestilence, but the threat of things putrid prevails, partly because some stenches are, literally, the scents of death and terminal disease.

Typhoid fever smells of freshly baked brown bread. Tuberculosis carries the sour scent of stale beer. Encounters with yellow fever are reminiscent of visits to a butcher's shop. The breath of some diabetic patients has the sharp odour of acetone. Sufferers from the rare, hereditary brain disease phenylketonuria are distinctly mousy. And even schizophrenia is accompanied by a characteristic sweaty smell. It still makes sense for diagnosticians to use their noses while waiting for the results of lab tests.

The nose often knows, even if the odour defies description. Death has its own smell. Every policeman and pathologist knows it well. It is a faint, sweetish, slightly faecal smell that, once experienced, you never forget or mistake for anything else. Even without injury or incontinence, human bodies begin to smell that way within minutes, while still warm.[40]

The chemicals responsible for this odour are soluble ones called indoles or skatoles, produced by the decomposition of proteins normally found in all animal intestines. But they occur also in oil of jasmine, and are used in the preparation of many perfumes.

There is no logical reason why decay would smell bad, why death should become more putrid than fragrant. But there may be a good biological reason.

Death is normal. It happens to us all, but the fact that it has happened somewhere near you, somewhere near enough to smell, could be important information. So it makes a distinctive announcement, giving out a warning you may not be able to see or hear, but can detect even at a distance, night or day, in any weather. And nothing does that better than a pungent, persistent and unavoidable smell.

There are hundreds of thousands of organisms which do not find the smells of decay and decomposition unpleasant. Bacteria, fungi, dung beetles, vultures and hyenas are positively attracted to anything putrid. For them, this is business as usual, dinner time. But even these necrophiles and scavengers have their limits. Each has its moment in the afterlife of a cadaver. Death is a banquet with a different set of guests for every course, a movable feast in which every diner knows exactly when to leave, moving on when the chemistry is no longer to their taste.

And it *is* a matter of taste, of being used to certain smells and finding some of them pleasant and others unpleasant. Some of this exercise is cerebral: we learn to like gorgonzola cheese despite its putrid smell. Some of it is chronological: we become more interested in musky smells and spicy flavours as we pass puberty. But some of it is purely biological: we are nauseated by the smell of decomposing flesh, because it tends to upset our stomachs. It makes us ill, because it makes us ill.

That is how aversion works. We develop an association between a consequence and a stimulus. If something has unpleasant results, then the steps leading to it come to be seen as unpleasant in their own right. There is a therapy for it, which most of us learn for ourselves. You sniff the milk in the refrigerator before drinking it; you sniff the air for signs of danger, and act accordingly.

In Texas, white-tailed deer at a feeding site were offered trays of their usual food pellets in the presence of fresh human blood or human urine. The deer were curious about the urine odour, but soon settled down to feeding close to it. The blood, however, startled them. As soon as the deer smelled it, they leaped back and from a distance began to stamp their feet in alarm, avoiding the area completely despite their hunger.[136]

Deer are hunted by humans in Texas, but have become used to feeding around the homes of their predators after dark. Human odours are now part of their everyday lives. But blood is not. Fresh blood, no matter to whom it belongs, is not normal – it is a dangerous smell, to be taken seriously and perhaps even genetically programmed as a source of alarm. Most mammals have no innate fear of carnivores in general. The smell of a European weasel, for instance, may put short-tailed voles on alert, but the odour of a jaguar leaves them unmoved. South American predators appear to have no olfactory clout in the British Isles.[183]

Anyone who has seen well-fed lions walking unchallenged through the herds in Ngorongoro Crater, will know that there is a strange accommodation between predators and their longtime prey. The relationship is finely adjusted, and ecological stability depends upon herbivores never becoming too good at escaping from a predator, and carnivores never becoming too successful at catching their prey.

Carnivores, on the whole, are extremely smelly. Their odours must be very familiar to prey animals, and often give the hunters away; but cat scents provide such a useful social and sexual service, that on balance they are worth keeping. And the gazelle and wildebeest simply learn to recognize when a predator with body odour is not dangerous, and when it would be prudent to raise a general alarm. Some species smell strongly only when they are alarmed.

As a child in Africa, I remember being taken to see the migration of a herd of springbok. These gazelles with short, curved horns once roamed the drylands at the edge of the Kalahari in groups so large they took days to pass by. It was like watching the whole desert move, as the river of warm brown bodies flowed across the plains, dividing to pass around a rocky outcrop, rejoining on the other side.

The seamless form of this great organism was interrupted only by an occasional individual animal, rising out of the whole, leaping vertically two metres into the air, paddling with its hind feet, raising a crest of snow-white hair on its rump so that it shone like a light. This performance is known as *pronking*, from the Afrikaans for 'showing off', but it is not just a dramatic display of high spirits and tight sinews. It attracts attention, perhaps even putting the individual at increased risk of predation. But it also serves another, more important function. The fringe of long hair on the rump straddles a fold of skin a little like a marsupial's pouch, but at the base of the erectile tuft is an array of sebaceous glands which produces a volatile odour that is common to all individuals in the group and elicits the same alert and evasive behaviour in the herd as a whole.

Ernest Thompson Seton, who hunted the prairies of Manitoba in the late nineteenth century, described a similar apparatus in the pronghorn antelope, when disturbed:

All the long white hairs of the rump-patch were raised with a jerk that made the patch flash in the sun. Each grazing antelope saw the flash, repeated it instantly, and raised its head to gaze in the direction toward which the first was looking. At the same time, I noticed on the wind a peculiar musky smell that certainly came from the antelope – and was no doubt an additional warning.[167]

He went on to say that he could smell the musky signal from thirty yards, and suggested that another antelope, given the right wind, could do so from miles away.

Mammals are naturally smelly. Most go around in the centre of an odour envelope, the size of which depends on their nature and their mood. Stress enlarges this envelope dramatically, spreading alarm among others of the same kind who may be nearby. And many rodents colour their envelopes at such times with an additive German researchers have called *Angstgeruch*, literally 'the smell of fear'.[134]

Such odours seem to be more than an emergency response. They linger. Frightened house mice leave traces of their concern so clearly on the surroundings that these paths of flight are avoided by other mice for up to eight hours. The major source of this message seems to be palmar glands on the bare skin of their feet.

This will strike a sympathetic note in anyone who under similar stress has found themselves with unpleasantly clammy hands.

The physiology, in humans and house mice, is much the same. Alarm triggers the production of adrenalin, the hormone that prepares the way for action – raising heart rate, blood pressure and blood sugar; dilating blood vessels serving the muscles and the brain, while contracting those of the surface skin; widening the pupils and making hair stand on end. All of which is designed to set an organism up for 'fight or flight', making it feel alert and look big, ready for almost anything. The fact that it also becomes sweaty is almost an afterthought, but an important one.

This is evolution at work. Few behavioural adaptations appear out of thin air; they are almost always based on existing anatomy and biochemistry. European minnows, for instance, have skin cells that produce a chemical which helps cuts to heal. The fish had such cells long before the release of the chemical after injury sent other minnows into the sort of dense school that represents their fright reaction. But the sequence is clear. First, there was the chemical; then the development of special cells (receptors) able to detect it; and finally, a measured response that turns the simple chemistry into a useful signal.[200]

So the tadpoles of common toads, which normally form orderly schools and swim in parallel rows, break up in confusion when they detect an extract of crushed tadpole containing the chemical bufotoxin. Their alarm response is purely olfactory. Tadpoles whose olfactory nerve has been cut show no response, not even when the rest of their shoal mates have scattered, swimming away into deeper water.[143]

Adult toads still produce the toxin, but they don't have to be mutilated or dead to release it into the environment. They make and store it in swollen parotid glands that lie just behind the eye. For them, the original skin balm has become a major chemical defence system, strong enough to paralyse or kill a predatory dog. But secretion of it still has the power to raise the alarm among other toads, even in the air and after surgery on their olfactory nerve, and they manage this because they now have a rival system of odour detection: they have Jacobson's Organ. As, of course, do we, which may make us susceptible in some surprising ways.

Weasels, civets and mongooses all have large anal glands that they can control at will, putting dabs of oily secretion on land-marks in their territories. New World skunks use the same glands in a more dramatic display. They jerk their bottom around to point directly at a potential predator and then raise their tails high in what is called the 'firing position'.[65] This is the first and only warning; what follows is awful.

A foul-smelling stream of noxious fluid is squirted out and up with extraordinary accuracy, directly at the face and eyes, from a distance of as much as six metres. The secretion is so foetid that it burns skin, wrecks the sense of smell for days and can never be removed altogether from clothing. Once 'skunked', twice very shy. It is the perfect deterrent and makes skunks highly self-confident little animals. But there is one mammal that has gone a step further, waging not just biological, but psychological war.

In the highlands of East Africa lives a rodent so peculiar that no one is sure whether to ally it with mice, rats or hamsters. So the crested whatever-it-is has been given a subfamily all its own. It deserves more – there is nothing quite like it anywhere else in the world. But there may be a message in its oddity that touches us all.

Lophiomys is about sixty centimetres long, shaped something like an American possum, but a lot more shaggy. Its entire body is covered with long, coarse, spongy hair that the naturalist Ivan Sanderson describes as looking 'like an old-fashioned shaving brush that has been soaked too long in hot soapy water'.[162]

Just like those brushes, the coat is badgery, composed of a mixture of black and silver hair that grows longer and straighter on the head and back, forming a crest that runs from the head to the tip of its tail. And all of this mane is erectile, standing up on demand to double the apparent size of this irascible rodent that turns its back on you, making an irritable growling sound if you even so much as look at it.

Along the flanks of *Lophiomys*, running all the way to its tail, are two belts of short grizzled fur that open away from each other like the pages of a book to expose a line of pale skin, pock-marked with waxy glands. When all this hair is standing on end, *Lophiomys* looks like an animated warning signal, striped light and dark like a skunk, presumably for the same threatening reason, but without the nauseating smell.

And there is more. On microscopic examination, every hair along the lateral band of glands turns out to be an *osmetrichium* – a 'smell hair', or wick. It is hollow and latticed like a loofah sponge, held together by a fine filigree of struts that encloses a space that becomes filled with glandular secretion. So when the visual signal of the raised crest is deployed, it comes with the addition of a potent olfactory wave diffusing from the extensive surfaces of the most complicated hair in the whole animal kingdom.

Lophiomys leads a quiet life, living in pairs in hollow trees or beneath rocks on the higher slopes of the Ruwenzori and Mendeba Mountains of Uganda and Ethiopia. There is nothing in these high-altitude habitats, no unusual concentration of owls, snakes, cats or other predators, to account for this animal's extraordinary battery of defensive weapons. So why the overkill? I don't know, but I can vouch for the fact that it works. In an area where all other mammals of this size form a prominent part of local people's diet, *Lophiomys* is never eaten.

As far as I can tell, without actually tasting it, the meat is not unsavoury and the lateral glands, though richly supplied with secretion, do not produce an offensive odour. But anyone who has been the focus of one of its bad-tempered tantrums will tell you the same story. This bristly little rodent has the power to make you feel dry-mouthed and distinctly uneasy, disinclined to have anything to do with it. And it seems to be able to do this with the help of a volatile, invisible mist which has no overt smell.

Lophiomys, it seems, has our number. It knows where we live and goes there, arousing primeval concerns. It evokes unreasonable unease by reaching out to parts of our brain that deal with such misgivings. This little merchant of the smell of fear bypasses the nose and strikes directly at the limbic regions of the brain. And that is very scary.

Everyone knows where the brain is. It sits right on top of the spinal cord in all vertebrates, taking input, creating output. Where we differ from other species, is in the size of the brain and in its connections.

Ours weighs on average nearly a kilogram and a half, and learns about the outside world through twelve pairs of cranial nerves that enter the skull through openings in the bone. The function of some of these is clear:

Cranial Nerve One is called the olfactory nerve, and carries information from the epithelium of sense cells in the nose.

Cranial Nerve Two is the optic nerve, and picks up news from the retina of the eye.

Cranial Nerve Eight is the acoustic nerve, bringing in signals from all parts of the ear.

So far, so good, but then things get complicated. Touch and taste are diffuse senses, gathering information in a variety of ways, most of them well known. But the true nature of smell, our most neglected and underrated sense, remains mysterious.

The olfactory nerve isn't our only source of news about odours. Cranial Nerve Five, the trigeminal nerve, is the largest of the twelve, gathering information from the entire face, including fibres from the mucous membranes of the nose and the sinuses. This seems to be an early warning system, protecting us from strong and harmful odours, to which we respond by pulling our head away as though we had been touched. The trigeminal nerve is usually thought of as part of our sense of touch, but it also carries signals from Jacobson's Organ.

So, with Cranial Nerve One picking up fragrant nose news, Cranial Nerve Five is free to deal largely with odourless phero-mones, and we have our two essential and parallel olfactory systems. It is clear, then, that smell in our species gets the sort of neural full-court press available to no other sense. And I suspect that overlaps between the two are frequent and functional.

It is hard to know what this might mean, but my guess is that it involves synesthesia, the combination of different ways of knowing about the same things. It seems to me that our noses do this all the time, taking up information in any way they can, mixing and matching as necessary, combining news, both volatile and odourless, in ways that give us powers still hard to describe.

I think that in this synesthetic harmony may lie the secrets of knowing things apparently unknowable, and recalling things with the sort of emotional clarity and immediacy that appears to be unavailable to everyday memory. In the search for meaning and the sort of utility that makes evolutionary sense, I am enthralled to discover that pheromonal activity is not confined to the animal kingdom. Plants have been picking hormonal news out of the air for a very long time.

There is, in natural history, a long-standing relationship between plants and the herbivores that eat them. As evolving animals develop new ways of exploiting vegetable foods, the plants keep pace with them, holding beastly appetites within reasonable bounds by inventing an array of new defensive mechanisms. The most obvious are physical deterrents such as heavy bark and sharp thorns, but there is also an impressive range of chemical constraints, similar to the one that clover uses to limit the number of sheep that graze upon it.

Tannins are a normal part of the defence system in many plants. They make leaves taste bitter and unpalatable, discouraging most animals from feeding on them. Tannins, however, are complex chemicals and expensive to make, so no plant can afford to manufacture them all the time. Most have arrived instead at a compromise solution, an accommodation which lets them produce these weapons when the need arises, but also allows herbivores to feed until the tannins have been brought up and put in place on the front line.

If you watch a browser in action, you will see that they feed on any one bush or tree for a relatively short period before moving on to the next one, even though apparently appetizing leaves remain within easy reach at the first. They don't go on eating at each site until all the leaves have gone, simply because the trees won't allow it.[204]

In South Africa, botanist Wouter van Hoven has discovered that acacias and other bushveld trees respond to browsing, or indeed to beating with a stick, by increasing the levels of tannin in their leaves within minutes. The response is most rapid and longest-lasting in times of drought, or when pressure on the plants is increased by having too many herbivores in a given area.

He also found, to his surprise, that tannin levels rose, sometimes by as much as 300 per cent, in neighbouring trees, even before the browsers or his beaters arrived. It seemed to him that the trees were able to raise an alarm and warn others of their kind of the danger of imminent attack. But he couldn't identify any mechanism for the pre-emptive response, and after it became clear that the whole idea caused alarm also among editors and referees on the more conservative botanical journals, van Hoven simply published his observations in a popular magazine produced by the South African National Parks Board. It appeared under the wonderfully provocative title of 'Trees' secret warning system against browsers'.[197]

The story went largely unnoticed until David Rhoades at the University of Washington in Seattle published his parallel study on red alders and Sitka willows under attack by insect pests. He found that leaves preyed upon by tent caterpillars showed a rapid increase of a chemical which slows the metabolism of such insect predators. But he also discovered that leaves from nearby control trees free of caterpillars showed the same response, but not samples from control trees some distance away.

Rhoades was unable to find any evidence of root contact between infested and uninfested willow trees, and ended his report with a point of punctuation seldom seen in learned papers. He wrote, 'This suggests that the results may be due to airborne pheromonal substances!'[155]

This suggestion, that plants may communicate with the aid of hormones in the air, is revolutionary. If trees can indeed warn each other of danger and respond to such warnings in appropriate and meaningful ways, there is little to separate them biologically from any animal that uses and reacts to social alarm signals. And Rhoades was not the only one thinking along such heretical lines.

Before long, he was joined by Ian Baldwin and Jack Schultz at Dartmouth College in New Hampshire. They worked with common poplar and sugar maple, tearing the leaves of seedlings and finding that this mechanical damage resulted in an increase of aromatic compounds called phenols, which are known to inhibit the growth of butterfly larvae. And the levels of phenol were as high in untouched control plants as they were in the experimental seedlings. The researchers also sampled air around the damaged plants and measured high concentrations of ethylene, a gas with a sweetish smell.[9]

We now know that ethylene is commonly released by plants under stress. And researchers at the Institute of Applied Physics in Bonn have found that an infrared laser produces an audible sound when exposed to the gas in air above tobacco plants deprived of water, or above geranium seedlings exposed to cold in transit to their markets. So it has become possible to hear the screams of plants in distress.[26]

At Rutgers University in New Jersey, tobacco plants deliberately exposed to tobacco mosaic virus have been found to produce a liquid known as oil of wintergreen or methyl salicylate. This is highly volatile and travels through the air to uninfected plants, where it is converted to salicylic acid, which provides some resistance to the disease. The authors who reported this study in the journal *Nature* in 1997 didn't exactly say so, but to me it looks very much like one plant giving another plant a pill to help it in a time of stress. Salicylic acid, after all, is the main ingredient in aspirin.[170]

So it is clear that airborne cues do stimulate changes in neighbouring plants and that trees can communicate with one another in very meaningful ways and at a distance. And now, Richard Hooley of the Institute for Arable Crops Research in England has announced that he and his colleagues are registering a patent on their discovery that rock cress plants have special receptors designed for responding to hormones in their environment. The interest of this group lies in the genetic manipulation of crop yields, but the truly awesome implication of their discovery is that the plant receptor is almost identical to the receptors in animal cells responsible for receiving chemical messages associated with taste and smell.[77] Hormones, it seems, were already in control long before plants and animals went their separate ways.

There is no longer any reason to deny the possibility that direct communication could take place between plants and animals through the action of pheromones. If one tree can warn another of danger, then why should we not share the real alarm of a giant redwood coming under the axe? 'Woodman, spare that tree!' Indeed. And suddenly, it no longer seems outrageous that truffles should share a pheromone with pigs. All living things have chemistry in common, and may well share certain kinds of chemical sensitivity. Forget about dogs sniffing trees – there are trees out there sniffing dogs!

Given their relative simplicity, and the fact that plants are incapable of running away, their response to being threatened is highly appropriate and very impressive. And it is interesting that their chemical of choice for sending out an alarm should be ethylene, known to us as an anaesthetic. Other plants respond to it as we do: they shut down. They close their stomata, stop sniffing the air around them and produce tannins. A sensitive mimosa, which normally takes avoiding action, cringing in response to cuts, burns, light touches or even changes in temperature, ceases to respond at all. But as the anaesthetic wears off, the plant slowly recovers and goes back into business again.

This, too, is appropriate behaviour. In danger, you do what you can and then relax. Some plants, when injured, even produce barbiturates which slow down all normal processes. Rice seedlings under stress secrete a substance similar to phenobarbital, which prevents normal sleep movements and conserves vital stores of energy until the problem has passed. The only problem is that these plants have the same difficulty with downers that we do – they can become as addicted to them as any human insomniac.[203]

Plants, of course, lack any true nervous system or a brain. But their capacity for meaningful communication, and the fact that some of them at least seem to be able to store and carry simple memories, suggests that the mechanics of information transmission and retrieval may be more fundamental than we now allow.

There have to be common denominators to a common chemical sense. Humans and dogs, bees and butterflies: all are repelled by the fumes of ammonia, and all are attracted to the smell of alcohol. They all dislike the bitter taste of quinine, and they all like the sweet flavour of sugar.

We all seem to have a unique body odour, but no one seriously suggests that each of the six billion humans now alive has a different body chemistry and a different odorous substance. The secret is in the mix, and there is an astonishingly large number of permutations of a small number of basic ingredients. Even the most complex genetic programs are written with just four molecules, made up of only sixteen chemical elements. Nature is notoriously niggardly about raw materials, and yet endlessly and generously inventive with them.

Most organisms find food, avoid enemies and locate mates in much the same manner: with chemistry. Even in plants, a simple chemical sense controls movement and growth. Roots reach out towards growth-promoting substances, and exude other chemicals which percolate menacingly through the surrounding soil, inhibiting the growth of rivals so well that many desert species become so evenly spaced that they seem to have been planted by hand.

And yet there is an amazing bond between all living things, something that seems to allow a potted plant, as found in one classic experiment, to become aware of the death nearby of brine shrimp being dumped at random intervals into boiling water.[203]

Why should this be? Even if dying shrimp do indeed send out a distress signal, why should it be of any interest to a rubber plant?

Birds, bees and trees all have alarm signals. And some of them are interspecific. Terns, plovers and willets, feeding in the same area as a group of gulls, all take flight at the sound of the gull alarm call. Such signals have high survival value and work well across the species line, but not all species function on the same frequencies or even with the same sense organs. So there could well be a strong natural pressure towards the evolution of a common signal, a sort of all-species SOS, warning perhaps of a common danger such as the approach of a seismic wave or the imminence of an earthquake.

A signal accessible to all life would have to be very basic. Almost certainly it would be something chemical, a molecular messenger such as the volatile hydrocarbon that sends mites of many kinds scattering out of an area when it blows through like a fire alarm. Or the stable nitrogen compound which works for all ticks, and goes both ways, stimulating dispersal when humidity is high, and encouraging aggregations in drier weather.[179] Returning to the question of individual odour, it seems that this may come about as a result of unique genetic influences on such simple substrates.

The luminous essayist Lewis Thomas suggested in 1974 that the secret of individuality in aroma may lie in immunology.[192] All vertebrates have a set of genes that direct the production of the proteins that are carried on the surface of every cell. These proteins enable our immune systems to determine whether a cell is our own, or belongs to someone else: self or non-self; to be nurtured, or to be destroyed. This set of genes is called the major histocompatibility complex and decides, for instance, whether populations of house mice which differ in just one gene can ever get on. Apparently not; a shift in a small part of only three out of more than four hundred amino acids is enough to drive a wedge between them.

Yet there are ways around the dilemma. Symbiosis happens: crabs and their commensal anemones; anemones and their damsel fish; ants and aphids all recognize each other as partners in special relationships by producing the necessary molecular badges and secret chemical handshakes. They don't treat each other as non-self, and it seems that this can happen because even very different species have a common ancestry.

The same chemicals, and the same components from which these chemicals are assembled, keep cropping up. The formula for the alarm signal of social insects such as ants and wasps is identical to the one their solitary ancestors once used for defence against a predator. And it remains one to which we are still programmed to react with distaste. It smells, quite strongly, of someone else's urine – hence the wonderfully expressive old word for an open anthill, a *pismire*.

There is a general and universal system of chemical communication in which all living things are involved. Each form of life, animal or vegetable, announces its existence through the Odornet, providing information on position and proximity, setting limits to encroachment or inviting participation in some kind of relationship. The result is a coordinated ecological mechanism for the regulation of who goes where, and how many can afford to do so. The whole thing seems to be designed specifically for keeping conditions on our planet within the narrow range acceptable for life, but it can be confusing.

As early as 1966, Harry Wiener at the New York Medical College wondered whether failure to fit into such a system might not be the cause of some of our most widespread but still poorly understood diseases. He proposed that there exists a common language, conveyed by external chemical messengers, in which conversation could take place between members of our species, without opening our mouths. He suggested that the alphabet for this language could be made up of individual chemicals. Words would be a mixture of such chemicals, and syntax provided by the intensity and frequency at which the signals were emitted. And he assumed that the stimuli provided by this language were ones which, for the most part, travelled through the central nervous system without alerting our consciousness.[209]

Wiener's initial proposal was straightforward. He was simply drawing attention to the existence of a system of unconscious communication by means of sending and receiving stimuli such as odours. House mice and humans certainly have such a working arrangement. But later, Wiener took his assumptions a step further. Normally, he said, we are not aware of these chemical cues. But suppose that someone were born with sufficient sensitivity to bring them, and the system, to conscious attention. What then?[210]

Wiener's own answer was simple. Someone like that, he pointed out, would be in trouble. He would be communicating in a non-verbal language that others did not seem to know. He would be able to perceive meanings, moods and intentions hidden to those around him. And if he were lucky and very well looked after, he might just be able to keep things quiet and make a good living as a perfumer or a priest. But more often than not, such a person would be ridiculed every time he tried to explain, or to convince others, of the reality of his insights. He would be unable to account for his own uniqueness, and would have feelings of doubt and persecution. He would be prone to panic and delusion. He would find it difficult to live with himself or others, and he would ultimately withdraw and deteriorate.[211] This condition, of course, has a name. We call it schizophrenia.

Wiener was right. Schizophrenia is the one disease in which abnormal messages seem to pass routinely from one individual to another. Therapists frequently find that diagnosed schizophrenics seem to be able to read other people's minds. Many have been disconcerted to find that such patients have 'an uncanny way of nosing out things about a physician's own personality, things which he may unconsciously have tried to cover up'.[83]

Some even describe a 'praecox feeling', a form of intuitive initial diagnosis evoked in a therapist on first meeting a schizophrenic patient. And because those with a heightened ability to smell anger, hate, fear or desire are also likely to have an increased capacity for producing their own odours, it should come as no surprise to discover that schizophrenics have their own characteristic odour. 'I can sometimes tell, by odour alone,' one colleague told Wiener, 'whether an entering schizophrenic patient is in crisis on that particular day.'[209]

The smell of schizophrenia is real. Dogs, and even white rats, have been trained to recognize it.[176] But it has proved difficult to find any metabolic source of the odour, which is said to be somewhat 'foetid'. That sounds like a pheromone, perhaps with a hint of bacterial decay to give it more body, something to make it, and the maker, more conspicuous to others with the same acuity. For those who are so different, there is comfort and survival value in finding others who share the same strange hypersensitivity.

Hallucination is said to be diagnostic of schizophrenia, a large number of patients complaining of unpleasant odours which no one else can smell. But, if Wiener is right, then it is possible that their perceptions are not delusional at all, but accurate descriptions of reality. There is a case on record of a patient who could detect the approach of his cousin, of whom he was fond, when she was several blocks away, still out of sight.[14]

Schizophrenics make most people uneasy. They offer contradictory and confusing information. They may seem to be indifferent, and yet put out odours which move us at an unconscious level. They induce awe and anxiety, a feeling not unlike that of meeting a half-trained tiger. This is the sort of feeling I associate with being close, in the wild or even in a zoo, to the crested and enigmatic *Lophiomys*. We are dealing here with a sense common to all life, with a sense that sniffs out strangeness.

The German philosopher Friedrich Nietzsche said:

What separates two people most profoundly is a different sense of cleanliness. What avails all decency and mutual usefulness and good will toward each other – in the end the fact remains: 'They cannot stand each other's smell!'[135]

After the French capitulation at Sedan in 1870, the Prussian army entered Metz, and everyone there held their noses as the triumphant troops marched by. The locals claimed later that five hundred cavalrymen barracked for just three weeks in Meurthe-et-Moselle, left thirty tonnes of excrement behind them. Over-eating, said the French, turns the skins of Germans into a third kidney.[113]

During the Great War of 1914, German soldiers in the trenches claimed to be able to tell if their opponents across no man's land were French or English by their smell. The English said the same, and called their foes 'krauts' from the perception that they lived on, and smelled of, sauerkraut.[113]

Jack Holly, a US Marine who led patrols behind the lines in Vietnam, said:

I am alive because of my nose. You couldn't see a camo bunker if it was right in front of you. But you can't camouflage smell. I could smell the North Vietnamese before hearing or seeing them. Their smell was not like ours, not Filipino, not South Vietnamese, either. If I smelled that again, I would know it.[62]

It is an unavoidable biological fact that human races have different smells and smell different to one another.

Half of the Vietnamese and Korean populations have no apocrine glands under their arms. In 10 per cent of Japanese armpits, the glands exist, but so sparsely they don't even come into contact with each other. And only 2 percent of Chinese have any underarm odour at all. So, to all Oriental noses, most Europeans and Africans, who are amply supplied with such glands, have a strong and disagreeable odour.

In addition, all developed nations have a tendency to look down their noses at less developed ones, most often finding fault with their hygiene. So Aborigines have been said to smell 'phosphoric', Congolese pygmies 'mousy', Hottentots 'of asa foetida', and the Caribs 'reminiscent of kennels'.[29] Sometimes, all that is necessary to draw the line between those who are like you and those of whom you disapprove, is to precede a racial or ethnic identification with the self-justifying adjective 'stinking'. The last word in any playground squabble is very often: 'And anyway, you stink!'[33]

Minorities very quickly become malodorous. In Poland, anti-Semitism was often expressed in terms of the smell of garlic. In the American South, white racists maintained social distance simply by persuading each other that black people were 'evil-smelling'. Class prejudices everywhere are supported by imputations that the lower class are 'foul-smelling' and must be avoided. In the 1930s, George Orwell said, 'The real secret of class distinction in the West can be summed up in four frightful words . . . *the lower classes smell.*'[137] And Somerset Maugham added, 'In the West we are divided from our fellows by our sense of smell. The working man is our master, inclined to rule us with an iron hand, but it cannot be denied that he stinks.'[122] Such stereotypes persist even in democratic societies, where rural people can still be excluded as 'mucky' or 'dirty'. And all these odious slurs are used, very effectively, to justify superior behaviour by 'good, clean, decent people'. Like ourselves.[111]

Part of this aversion has to do with familiarity. At the University of Tsukuba in Japan, subjects of similar ages and comparable social backgrounds from Germany and Japan were offered a wide variety of smells to assess for pleasantness. The three smells the Germans found most unpleasant were all Japanese – fermented soybean, dried bonito flakes and cypress oil. And the three smells the Japanese most disliked were all German – blue cheese, strong sausage and church incense. And this sort of xenophobia manifests itself just as powerfully when it comes to personal experience.[7]

We obviously use smell as one of our most basic markers when meeting and assessing strangers. Anyone who doesn't conform with the olfactory norm is either an outsider or not to be trusted. And no amount of aromatic subterfuge can quite confound our noses, because this norm is one established at a subliminal, unconscious level, often via Jacobson's Organ. Jean-Paul Sartre, in his assessment of the sensuous French poet Charles Baudelaire, said: 'When we smell another body, it is that body itself that we are breathing in through our mouth and nose, that we possess instantly, as it were in its most secret substance, its very nature.'[163]

Exactly. And not just in romantic passion, but always, unknowingly, padding out what we know of others by a clandestine olfactory probe into their best-kept secrets. As a result, we take a calculated chance with an outsider about whom we know very little, or take an instant and unreasonable dislike to someone to whom we have just been introduced by an old friend. The nose is the body's advance guard. It continues to scent out the truth. Without it, we are in real trouble.

There is an enormous variety of olfactory sensitivity within our species. In any smell test, results can vary by as much as a thousand-fold. Anything more than that is considered to be abnormal and worthy of special mention. Sufferers from cystic fibrosis, who sweat enormously, also have the ability to detect some smells at levels ten thousand times weaker than the usual threshold. Most variations, however, are in the other direction, towards lower sensitivity.

Loss of smell is called *anosmia*, and is believed to come in three forms. In specific anosmia there is an odour blindness to single substances. General anosmia occurs when the olfactory nerve is injured, or there is a loss of sensitivity in the membrane of the nose. And complete anosmia involves the loss also of Jacobson's Organ. But there are few real experts anywhere on any of these disorders. They are hard to treat and there is no clear consensus, even among the experts, about what any kind of smell deficit may mean to our species.

Those who cannot see are blind. Those who cannot hear, deaf. Those who cannot speak, dumb. But those who cannot smell are left hanging: they suffer from an absence without a name. Smell-no-evil is a condition that doesn't even deserve a monkey.

The result of losing the nasal part of your sense of smell is often deep depression. Taste goes with smell, and all that is left of gourmet cooking is the texture and temperature of the food. Those who suffer such a loss suddenly say that it is like forgetting how to breathe. We take smell for granted, unaware that everything around us has a smell – until, that is, nothing smells at all. Then life loses a lot of its succulence, and hardly feels like living any more. There is a high suicide rate among anosmics. Those who do hang on find it increasingly dangerous to do so, for without smell there is no ability to detect the smell of smoke or gas leaks, or the odour of putrid or poisonous food. But no one knows for certain what the consequences are of losing your Jacobson's Organ. If humans are anything like the mammals used in olfactory research, then the results could be devastating.

Without Jacobson's Organ it is possible that normal reproductive physiology will be impaired, and that puberty could be delayed.

Without direct access to the usual pheromonal information from your own and the opposite sex, it may be difficult for your body to make the proper responses.

Without signals from Jacobson's Organ, the limbic region of the brain may fail to initiate fight, flight or avoidance mechanisms appropriate to the occasion.

Without the necessary endocrinal control, there could be major and inappropriate swings of mood. With no subliminal messages, it would certainly be more difficult to assess the intentions of others. And with limited access to the non-verbal communication going on all around you, the chances are high that you will become paranoid and delusional. And there is very good reason to take such concerns seriously, because there is a growing population out there who have already been deprived of their Jacobson's Organ.

It is difficult to assess the numbers of nasal operations performed each year, but one conservative estimate is at least a hundred thousand.[60] Some of these are benign, but a good many are not. Every time a surgeon slices away at a nasal septum in the name of fashion or vanity, both sides of Jacobson's Organ are at risk of being damaged or even removed entirely, without thought for the consequences. This is simply because in most medical circles the textbooks in use ignore the organ altogether, or treat it as a vestigial structure, something dispensable, without function or purpose.

That is unnecessary and probably unethical. There are techniques which allow the mucosal lining of the nose to be preserved, even in rhinoplasty or work on septal deviations. And such measures should be obligatory, at least until we know more about how the Organ of Jacobson works in humans. So be warned! If you are considering cosmetic surgery on your nose, know that it could deprive you of the very things in life which having a new, cute, little button nose were supposed to improve.

Aromaticos

To complement 'fragrant' sources, Linnaeus created 'aromatic' ones from the Greek aroma – which encompasses the subtle, persuasive smells of all the spices. These have an essence whose nobility is best released through smoke and fire, per fumar, producing perfumes.

'That which is set on fire' is incensed, offered to a god to flatter or inflame, turning gums, resins and essential oils into aromal breezes of the sort that might have pleased even Apollo.

'To spice' means to season or perfume, but a spice is also a species, a kind of citron, anise, cinnamon or clove, waiting, in Alfred Tennyson's words, 'to feed with summer spice the humming air'.

From victorious athletes to poets adorned with laurel wreaths, aromatic substances suffuse, adorn and crown our endeavours. In 1825 Michael Faraday discovered benzene in the gas rising from burning whale-oil lamps. Forty years later, the German chemist August Kekulé dreamt one night of a snake in whirling motion, swallowing its own tail. This vision inspired the work in which he not only solved the cyclic structure of benzene and other aromatic hydrocarbons, but also provided a theoretical basis for the whole of organic chemistry.

6

Smelling good

Stench was once synonymous with disease. It is hard to imagine how crowded cities in the eighteenth century must have smelled. But French historian Alain Corbin has succeeded in capturing the essence of the time in a superb piece of olfactory archaeology called *The Foul and the Fragrant*.[33] In it, he conjures up a frightening brew of miasmas, emanations and malodours: rising up from cesspools in the cellars of each building; seeping into every part of the fabric and every aspect of daily life; overflowing into the streets in floods and hot weather, bringing life to a standstill each time a cesspit needed to be cleared of the worst of its effluvia. Small wonder that such smells filled, not just the nose, but also the mind.

In both France and England at that time, there was much talk about the smells of decomposition and the 'odours of corruption', leading to anxieties new to our species. All at once, we were beset by 'morbific vapours', 'deflagrations' and 'stinking effluvia'.[33] And the air was full, too, of terrible stories of the mishaps that befell cesspool cleaners; and the excremental fate of lost travellers, swallowed up in the sewers beneath the cities.

Such concerns had two direct results. The first was a flurry of scientific activity in which Joseph Priestley and Antoine Lavoisier identified oxygen, Robert Boyle gave air an official pressure and substance, which Louis Pasteur later populated with microorganisms which could be identified and controlled.

The second effect was a social one, arising from the realization that if air was simply a transport system, carrying particles of dust, smoke and the essence of food and faeces, then it could also carry human information.[156] On the eve of the French Revolution, Tiphaigne de la Roche announced:

Particles of an invisible substance called *sympathetic matter* spread around men and women. These particles act on our senses, and this action produces attraction or aversion, sympathy and antipathy. With the result that when the sympathetic matter which spreads around a woman, makes a pleasing impression on the man's senses, thenceforth that woman is loved by that man.[33]

Which nicely anticipates the idea of pheromones by two full centuries.

So, out of stench grew new ideas about desire, making odours attractive enough to catch the attention of amorous adventurers such as Giovanni Giacomo Casanova. They fired the imagination of Pierre Cabanis, the father of modern physiological psychology, who in 1802 christened our sense of smell, 'the sense of sympathy'. And before long Johann Goethe was confessing to the theft of one of Madame von Stein's bodices so that he could breathe it at leisure. The novelist Restif de la Bretonne was inventing leather fetishism by sniffing the shoes of chambermaids, and his contemporary Barbery d'Aurevilly had one of his heroines seduce her *abbé* by sending the defenceless man one of her well-used chemises.

All of a sudden, body odour was fashionable. Even in a time when a Parisian bureaucrat was complaining that, 'The capital is nothing more than a vast cesspool, the air is putrid and so foul that the inhabitants sometimes can hardly breathe.'[33] The image of young girls strolling past corpses piled up in the Cemetery of the Innocents, buying fashions and ribbons despite the cadaverous odour, is a resonant one. The intolerable becomes tolerable in time. Our threshold for every smell is a movable one, and it rises, not so much because sense cells in the nose become exhausted, but because the brain gets bored with even the most alarming odours. But that, too, can be dangerous.

If you put a frog into a pan of warm water, it will try to escape. But if you heat the water very slowly, the frog notices nothing and allows itself to be boiled to death.

There are few chances to study human behaviour in the wild, away from ethnic and national influences, without any cultural baggage. We cannot just isolate children from birth to see what happens. But on a few occasions, such isolation has taken place anyway.

At the turn of the eighteenth century, a boy who seemed to be about twelve years old was found running wild in the woods of Aveyron in the Massif Central of France. He was captured and taken to Paris, christened Victor, and examined by a number of scientists before his death there in 1828.[110]

In the same year that Victor died, another boy was discovered, apparently recently escaped from the confines of a cellar near Nuremberg in Southern Germany. He appeared to be about sixteen years old, carried a note giving his name as Kaspar Hauser, and lived in care for five years until he was murdered, in 1833, apparently as part of a political intrigue that grew around rumours of his royal origins.[175]

And in 1920, two girls aged about eight and two years old were found living in a den in India with a she-wolf and her cubs. The younger child died soon after she was rescued, but the older girl was named Kamala and survived for a further nine years in an orphanage.[116]

These examples give us a glimpse into the way senses develop outside culture, and it is interesting that all three children had extraordinary olfactory abilities.

Victor, who grew up in the woods, smelled everything available, apparently with huge enjoyment. He found excitement even in pebbles that appeared odourless to anyone else. He was extremely sensuous, rolling in the snow with delight, and drinking nothing but fresh water, which he savoured 'as though it were an exquisite wine'.[29]

Kamala, the wolf-girl, preferred raw meat, whose odour she could detect at a great distance. She liked milk, but never drank water. And she was so indifferent to cold that she had to be conditioned into feeling differences in temperature.

But Kaspar, who had reputedly lived in the dark and been deprived of everything but bread and water, had the most prodigious senses of all. He could distinguish between fruit trees at a distance, with his eyes shut, just by the smell of their leaves. He could tell people apart by the sound of their footsteps. He could feel the difference between metals by touching them with his fingertips, and distinguish between the north and south poles of a magnet. But in the real world he seemed helpless, overwhelmed to the point of nausea by the avalanche of stimuli, of colours and aromas, available to him there.

It seems, from the reactions of all three of these 'wild children', that the senses remain relatively unimpaired by isolation, though they tend also to lack regulation and an appropriate sensory model. Kamala was the only one of them to have been instructed by another living thing, and she was distinctly lupine in her preferences.

It seems obvious, too, that smell, more than any other sense, provided each child with what little self-identity it could muster. Kamala used her nose to search for traces of her young companion after the death and burial of the child. Victor was unable to recognize and accept new acquaintances until he was allowed to sniff their hands and cheeks to assure himself of their identity. Kaspar's pain and panic in a world full of odours persisted until such olfactory dominance could be tempered by dilution with social skills.

That, in the end, was the message of the feral children. Culture dulled all their senses, according to its own ranking of the faculties. Those looking after Victor, for instance, dismissed his olfactory enthusiasm as a 'primitive' talent, and spent their time trying to interest him instead in 'the dainties of French cuisine'.

All our senses are more or less reconstructed according to the prevailing social programme. But it is heartening to learn that, despite everything Victor's mentors could contrive in twenty-eight years of acculturation, there was little they could do about his Jacobson's Organ. He could be persuaded to accept cooked instead of raw meat, but he could never be taught to curb his sexual impulses, making fervent advances to any female who appealed to him.

There is another uncultured individual in the olfactory literature, a fictional figure, but one with a revealing talent. In *Perfume*, Patrick Süskind has followed the Hollywood tradition in creating a biological anomaly: a hero who really has no body odour. He doesn't smell at all, but of course smells extraordinarily well. It makes a nice change from X-ray vision.[188]

Grenouille is an orphan, a feral child who educates himself, learning the words of smell sources first – fish, cabbage, goat, and so on. Then all the flavours of milk, all the tastes of smoke. He navigates in the dark:

With his eyes closed, his mouth half open and nostrils flaring wide, quiet as a feeding pike in a great, dark, slowly moving current. And when at last a puff of air would toss a delicate thread of scent his way, he would lunge at it and not let it go.[188]

As he grows, he feeds upon the invisible gruel of Parisian odours, cooking up new scents in the olfactory kitchen of his imagination, learning the language of perfumery and becoming, before long, the best nose in the world, able to 'snatch the scented soul from matter'. But, in addition to his lack of odour, he also lacks a moral conscience and begins to build himself an artificial personal smell, by distilling it from murdered maidens.

And he succeeds all too well, investing himself with an odour so seductive he is torn to pieces and devoured by a smell-crazed crowd. But the insights of the story are profound. Grenouille's supersensitivity to smell makes it possible for him to 'see' through walls, sniffing out who is in a closed room; tell which cabbage has a worm in it; predict the arrival of a visitor or the approach of a storm. He produces all the evidence of what we call 'second sight' by the exercise of what we might instead call 'first smell'.

If we had even 10 per cent of the olfactory acuity of a dachshund, we could do as well. Smell is the witch's sense, sniffing out the spirit of what has been, detecting an essence after the fact of its existence. It is the formula for time travel, lingering on for decades as the scent of cedar in an old sea-chest. It lies at the heart of premonition and clairvoyance, carrying our consciousness well outside the confines of the body.

Perhaps we can already do great things. I believe that where *Perfume* ultimately fails is in its emphasis on the dangerous savagery inherent in the sense of smell. Süskind lingers on the details of fragrant, hapless maidens stalked by an obsessive maniac who sniffs out his prey, instead of exploring everyone else's untouched potential. This is a little disheartening. It really is time we stopped denigrating the power and influence of smell in all our lives. And yet, smell has been culturally suppressed for so long that one has to wonder. Perhaps heightened olfactory consciousness really would be dangerous to the established social order. Think what we could do with it!

For a start, we would finally get to know who the good guys are. Historically, the gods always smelled good. Zeus was described as 'wreathed in a fragrant cloud'. Aphrodite left all of Cyprus sweet-smelling behind her. Venus was 'perfumed with the rich treasures of the revolving seasons', but her casual gift of some of this essence to a mere mortal proved fatal: he was killed by a jealous husband. Hades enticed Persephone with a scent 'so sweet that all heaven and earth laughed with pleasure'. Hera seduced Zeus himself by 'cleansing from her lovely body every stain, and anointing it richly with oil, ambrosial, soft and of rich fragrance'. And Mount Olympus stood at the centre of all this heady perfume.[192]

Given all of this, it is hard to imagine why we should ever have believed that what the gods really wanted was the stench of burnt flesh from sacrificial animals. Happily, that fashion passed and was replaced by one in which a god's attention was courted instead by offerings of fragrant incense – 'that which is set on fire' – in place of a living sacrifice. And interest turned finally to a specific smell, a mystical fragrance believed to adorn an individual blessed with divine favour: to a search for the 'odour of sanctity'.

Saint Paul said of his priests, 'we are the aroma of Christ' – a quality imitated by hopeful clerics with rose garlands and censers on feast days but, it is said, worn naturally by those imbued with real grace. There was a competitive quality to the fragrance bearers. Saint Lyddwyne, we are told, boasted seven distinct flavours of sanctity. Padre Pio could muster six, and Santa Teresa, in all her usual modesty, managed but four. All are best re-membered, however, for the odour of their going. Saint Simeon's death was accompanied by 'an incomparably sweet fragrance'. When Saint Patrick died, 'a sweet aroma filled the whole room where he lay'. And when Saint Hubert died, all Brittany was reputedly suffused with his fragrance.

Conversely, sinners can be recognized by their stench. If the good smell good, then the bad must smell terrible, which is but a short step from 'those who smell terrible are obviously bad'. But there may be some basic biology underlying both assessments. The clue lies in the word *inspire*, which covers both breathing in and infusing with feeling, making *inspiration* something emanat-ing from an influential, perhaps strongly olfactory, person. Some-thing, moreover, which, with the help of Jacobson's Organ, can go directly to the deepest levels of the unconscious mind.

Nothing cuts through the surface of things like an odour. And it works coming and going, bringing in news and carrying out an essence of a person's soul. It is no accident that Venus, the

Madonna, the heart, passion and the angels are all symbolized by the rose. Remains of roses have been found in Egyptian tombs. Romans scattered their petals at banquets, threw them in the paths of victors, and drank rose oil in their wine. This 'Queen of the Flowers', whose scent is described as rich, sweet, tender and warm, is venerated also in Christian, Rosicrucian and Sufi traditions. The Hindu word *aytar* describes a mixture of sandalwood and rose oil, the 'attar of roses' widely used in ritual worship. And it was the mainstay of monastery gardens in Europe during the odorous Middle Ages. Small wonder, then, that recent clinical studies have found that rose oil acts as a mild sedative and antidepressant, soothing nervous tension, slowing heartbeat, lowering blood pressure and increasing concentration.[112]

On the other hand, bad or astringent smells, things like burning flowers of sulphur, are associated more often with *fumigation* – another loaded word, originating in the exorcism of demons by smoke too thick to let even the denizens of brimstone breathe. The philosophy of the demonifuge thrives still in the strange world of television commercials where household cleansers are touted for their germicidal properties, but sold largely on the strength of their scent.

To be clean, it seems, things have to smell, preferably of lemon or pine, both of which happen to have natural insecticidal properties. These have unfortunately long since been supplanted by less expensive artificial substitutes, that no longer attack bed mites or '99 per cent of all known germs' on sight, but continue to move little-known actors to ecstasy.

Pure pine oil, with a scent that has been described as penetrating and turpenic, actually does repel lice and fleas. It fights respiratory infections, cleans the sinuses, is antirheumatic and stimulates the adrenal cortex to produce steroids that suppress allergies and reduce inflammation. In its purest form, the oil of Norway spruce in a sauna aids even in the elimination of toxins through the skin. Demons, beware!

Devils may leave through the skin, but magic most often chooses to enter through the nose. So in Europe, a sprig of rosemary beneath the pillow is thought to protect a sleeper from nightmares. And sprays of St John's wort are still hung above the doors of houses at the summer solstice. But the most effective method of invoking magic everywhere seems to have been to heat herbs and essences in special vessels on censers. And anything treated in this way has come to be called 'incense'.

The earliest recorded use of incense comes from China, along with cassia, cinnamon and sandalwood. And the practice spread through the Hindu civilizations, which added frankincense, lime and jasmine to the recipe. But it was the Egyptians who seem to have formalized the production of aromatics, introducing myrrh, laudanum, galbanum and styrax to the repertoire, and turning it into a cult. Incense in Egypt was a total spiritual experience. It purified and protected believers, and acted as a channel between them and their gods. Nefertum, the god of perfumes, could be reached only by the prayer line of scented smoke rising from a censer. Ramses III alone was said to have used almost two million jars of incense during his reign, most of it going to the Theban temple of the god Amun, known to the Greeks as Zeus.

The Greeks learned of incense from the Egyptians in the seventh century BC, and soon no ritual in the Peloponnese was complete without frankincense or myrrh. The heaviest use was probably that made in the frenzied Dionysian rites of communal ecstasy, which overflowed later into the Bacchanalian revels of Rome, and the excesses of Herod. So the natural response of the early Christians was to take a long step back and wonder, 'What is going on here? Do the gods really have nostrils?'

A fair question, but one which misses the point. Incense is a human invention, intended for human use and the human nose. And the most interesting feature of incense, no matter who makes it, is that it comes from five principal sources: myrrh, frankincense, laudanum, galbanum and styrax. These are all resins – known respectively as balsam, olivanum, onycha, asa foetida and balm of Gilead – taken from desert shrubs in Arabia and North Africa.

Resin is a natural biological substance. It is produced by trees which have been injured, and oozes from the bark to form an antibiotic scab that protects the plant from infection by bacteria, fungi and other pathogens. These substances have large molecules, which give resin its viscous quality. But what matters most is that all these ingredients of incense contain resin alcohols, called phytosterols, which biochemically are remarkably similar to human hormones: notably the ones produced in our armpits, carried on our breath and excreted in our urine.

This similarity has not gone unnoticed. The poet voluptuary Robert Herrick makes no bones about it in his lines 'Upon Julia's Unlacing Herself':

> Tell, if thou canst, and truly, whence doth come
> This camphire, storax, spikenard, galbanum,
> These musks, these ambers and those other smells

Sweet as the vestry of the oracles.
I'll tell thee: while my Julia did unlace
Her silken bodice, but a breathing space:
The passive air such odour then assumed,
As when to Jove great Juno goes perfumed,
Whose pure immortal body doth transmit
A scent that fills both heaven and earth with it.[120]

His delight and ecstasy in Julia's body odours is obvious, and it is highly significant that his olfactory allusions are not to any of the familiar perfumes of the day, but to named incenses most often encountered in another context altogether.

The Austrian perfumer Paul Jellinek put the five main ingredients of incense to a more objective test, by asking a panel of professional scent-testers to tell him if they bore any resemblance to body odours. There was wide agreement that myrrh, which has a somewhat sour smell, is very like the underarm odour of fair-headed people. Frankincense, which has a sweet smell, more closely resembles the armpit odours of dark-headed people. The balsamic laudanum is reminiscent of the smell of head hair, and the sweaty styrax of general skin odour. And to clinch the connection, the panel tried mixing the incense ingredients in with a basic floral eau de Cologne, and found that in every case this made the product more erogenic.[87]

It is notoriously difficult to separate innate responses to such pheromonal material from learned responses to the odours of incense or any other fragrance. We could find some smells exciting, or even sexually stimulating, simply because of previous personal experiences associated with them. A lot of precisely controlled work still needs to be done, but it is at least intriguing to note that the herbs we choose to burn on our altars are not chosen at random from nature's vast pharmacopoeia. They are all rare and expensive resins, chemically very closely related to human steroids.

The ingredients of incense smell remarkably like those produced on our skin and in our urine. As far as we can tell, they are perceived by all of us, both consciously and unconsciously, in the same way and along the same sensory pathways as human sex hormones. And once vaporized over hot coals, they produce a thick white smoke which contains precisely the sort of heavy particles most likely to be picked up by Jacobson's Organ and to go straight to the emotive limbic areas of our brains. They are unquestionably exciting.

It also needs to be said that incense is a very good example of how odour crosses ethnic lines, uniting people in an experience common to all members of our species, instead of isolating us along racial or ethnic lines. One of the consequences of this is that we are more likely, in the presence of incense, to experience the sort of communal ecstatic feelings on which organized religion depends for success. If there is a 'God gene', which predisposes us to accept the idea of a deity, it may be sparked into action by the kind of basic biological euphoria that incense stirs in us.

I am surprised that news of this has not got out. Michael Stoddart has made a very persuasive case for it in his book *The Scented Ape*. And by now the *National Enquirer* ought to have run a front-page story under the headline SEX IN CHURCH. My guess is that no one wants to be involved in the sort of outrage which is bound to greet any attempt to suggest that the pleasure we find in religious ritual has anything to do with the smell of sex. But it is hard to explain the global popularity of these particular resins for these particular occasions in any other way.

Incense needs a censer, a source of heat to release a fragrance, but there is another way of getting an aroma around: by using the warmth of our own bodies to propel it. Perfumes, as their name implies, began life as ritual sources of odour, spread by heat and smoke. But as far back as ten thousand years ago, they became far more personal. Credit, as usual, goes to the Chinese, whose legendary Yellow Emperor is said to have introduced perfume to the world. In India they give fragrant precedence to the god Indra, who is usually depicted riding on a white elephant, a male elephant in musth.

Early perfumes were solid, little more than scented grease, mostly animal fats or unguents charged with soaked herbs. But five thousand years ago, in the Indus Valley, terracotta stills were being used to extract essential oils and preserve them in alcohol for later use. Two millennia after that, Egyptians were mass-producing scents they called 'the fragrances of the gods' in work-shops in Thebes, and exporting these trade goods to Mycenae and Assyria.

To begin with there was certainly something sacred about perfume. Its connection to the 'odour of sanctity' is perhaps that the first smell to follow death, before putrefaction, is sweetish, fragrant enough to be associated with the departure of the soul. To embellish this connection, in all parts of the ancient world the dead were anointed with essential oils, washed in the extracts of fragrant woods, covered in flowers and fruits and buried with consecrated perfumes.

Hindus wash their sacred images daily with musk. Buddhists cleanse their statues with aromatic oils. Perfumes were offered to the Mycenean gods, and Egyptians anointed statues of their gods with 'ambrosia' or oil of myrrh. And from there it was a simple step to consecrating priests and kings in the same way, creating a fragrant social distinction which still exists. The rich *are* different: they can afford to smell better.

So perfume starts as something magico-religious, a symbol of transformation that is confined to sacred ritual. But everywhere it develops rapidly into something far more profane as the secret spreads from the priests to the people. This transition is easy in the Orient, where aromatics have always played an overt part in religious ecstasy. In Tantric ritual, sandalwood oil is applied to a man's forehead, chest, underarms, navel and groin. And a woman is similarly anointed with jasmine on her hands, patchouli at her neck, amber over her breasts, musk in her groin and saffron on her feet. A wonderfully heady mix of spiritual realization and pure sex.

In the West, fragrant obsessions tend to begin instead with conspicuous consumption. Cleopatra goes to meet the emissary from Rome steeped in jasmine, sandalwood and olive oil, decorated with henna and kohl, travelling on a barge whose sails are soaked in rose water, and on whose decks huge burners are piled high with *kyphi* – a concoction of cinnamon, pistacia, juniper, cyprus, honey, frankincense and myrrh. 'Even the winds,' said Plutarch, 'were love-sick.'[144] Mark Antony didn't stand a chance.

The Romans were quick to see the point. In the capital, the cult of perfume became an important part of a hedonistic lifestyle. And at the funeral of his wife Poppaea, Nero burned ten tonnes of incense to perfume not just the air, but the trees and every mourner in Rome. But that wasn't just an imperial prerogative – everyone wore scent of some kind. In Capua, a whole street was given over to perfumers, and even women of modest means employed *cosmetae*, slaves whose sole function was to keep their mistress fragrant. To this end they bathed her in water strewn with the leaves and flowers of a herb which took its name from the Latin *lavanda*, for 'one who washes'.

If it smells good, it must be clean. And conversely, if it doesn't smell at all, it can't be any good. Which is why odourless disinfectants are routinely marketed with smells sharp enough to take your breath away. The medieval notion that strong smells are effective as a protection against pestilence, dies hard. Hence the laurel wreaths on the heads of Greek heroes; bags of herbs that are still hung around country homes in modern Greece; cloves of garlic worn around the neck in Italy and Romania; and the *lei* garlands given to visitors in Polynesia that continue to display a fine ambivalence between hospitality and hostility, something midway between 'welcome' and 'go away'.

From the fall of the Roman Empire until the Renaissance, perfumery was in Arab hands. It was in the Middle East that the techniques of maceration, enfleurage and distillation – soaking, compressing and boiling fragrances out of plants – were refined. And it was there too, that the three levels of perfume harmony were refined.

The creation of a successful new fragrance requires 'top notes' – usually taken from the sexual secretions of flowers such as magnolia, wisteria and some orchids, which contain oils designed to attract pollinators, often by imitating their own sex attractants. It is even possible that some insects, such as bees, copy the plants and use flower fragrances as chemical precursors for their own hormones – a nice piece of reciprocity. The 'middle notes' in a perfume are provided largely by the classical resins of galbanum, frankincense and myrrh. These happen, no one knows why, to resemble the sex steroids of mammals. And the 'base notes', which are supposed to fix a perfume, are also very much like mammalian pheromones, this time borrowed from the sex glands of civet, beaver and musk deer, which have a distinctly urinous or even faecal odour.

In short, to make a perfume that appeals to our species, you have to go right back to basics and mix in a little of whatever works for moths and magnolias, parsnips and pigs, when each is playing the mating game. You sell your fragrances on their floral notes because, once again no one knows why, we are as strongly attracted to roses and violets as is any bee. And you tailor them to specific markets, which appear to coincide with the hair colour and complexion of women. Surveys suggest that blondes prefer fresh, stimulating odours, like those of mimosa and hawthorn. Redheads go for exciting smells, perhaps orange blossom or honeysuckle. And raven-haired ladies prefer something more sultry, more like orchids and magnolias. Brunettes are difficult, with preferences ranging all the way from soothing lavender to intoxicating violets.[87]

A more objective study carried out in Germany found correlations between recognized personality types and their odour preferences. Extrovert perfume users – sociable, cheerful, impulsive people – showed a tendency to choose 'fresh' odours such as Eau de Courrèges or Eau de Guerlain. Introvert perfume users – reserved, serious and careful people – had a strong preference for 'oriental' odours such as Shalimar or Opium. And emotionally ambivalent users played safe with simple 'floral' fragrances such as Rive Gauche and Miss Dior.[123] These were tests in which the subject knew nothing of their psychological assessment and were never exposed to the perfumes' names or brand images. And yet their choices show an interesting pattern of 'fragrance fit', each person going for an odour signature that seems to identify them better than most people's passport photos.

If the purpose of perfume is simply to mask our own body odour, a way of preserving olfactory anonymity in public, then anything unfamiliar would do. Or natural selection could by now have come up with a range of home-grown masking substances, manufactured in the body and deployed as necessary as deterrents. But that seems, on the whole, not to have happened.

In the absence of Napoleon, who preferred and praised her own exciting natural fragrance, the Empress Josephine is said to have concealed her charms under so much musk that the servants in her boudoir frequently fainted. Very strong scents are repulsive to both sexes, but there are obvious dangers in such smelly subterfuge, which sends the sort of mixed message that can easily backfire. So, over time, those who use perfume learn to choose scents that are less ambivalent. Most purchasers defend their choice by saying simply, 'I like it.' But the fact that there is sufficient consistency in such choices to successfully allow the manufacturers of Opium to target introverts, and of Guerlain to appeal to more extrovert users, suggests that there are biological factors at work here, as well as shrewd marketing.[212]

We seem, by and large, to have an unconscious ability to choose well, to purchase perfume that complements our own biochemistry in ways that provide an appropriate olfactory passport – one that gets us through social and cultural checkpoints to where we need to be. Those who make bad choices, who fail to pass olfactory inspection, are painfully obvious, as conspicuous as those who dress with poor colour sense.

It seems clear that the fragrant choices we make are largely unconscious ones, governed not by the high-note banner headlines of each perfume, but by the message hidden away in the fine print at its baseline. It is written there in a chemical code that only the unconscious brain can read, because that is how it all began, back when it was necessary for women to be secretive about their reproductive condition, lest ovulation become attractive to the wrong men, or menstruation become disturbing to the right prey animals.

The only certainty in all this is that our perfumes of choice, all the best-selling, best-known, most popular fragrances in the world, are pheromonal. From all the hundreds of thousands of natural sources of scent, no matter how rare or how expensive these might be, we have settled on a strangely short list of vital ingredients. And every one of these carries the same potent, and mostly subliminal, message. Perfumes may have become part of sacred and social ritual, but at heart they are about just one thing – sex. This may explain why we seem to be ambivalent about perfumes, as we are about smell in general.

In truth, we are a smelly species. We have dense patches of scent-producing glands in our armpits and around our genitals, and in both these areas we have retained tufts of hair that are very conspicuous on our otherwise naked bodies. We are better equipped for smell-signalling than any of our nearest relatives, but seem hell-bent on denying the fact.

It is not that we cannot smell ourselves. That is one truism that is patently false, even without Jonathan Miller's description (in 'Beyond the Fringe') of our species as one prone to doing 'those little personal things people sometimes do when they think they are alone in railway carriages: things like smelling their own armpits'. We are, in fact, acutely aware of our body odour at all times, and spend an extraordinary amount of time and energy on removing it, and then replacing it with something else – something equally odorous and obviously sexy, unconsciously drawing attention to the very thing we consciously try to hide.

This sort of ambiguous behaviour is usually a sign of repression. Sigmund Freud blamed sex for most of our emotional problems and neuroses. He believed that the greatest conflicts arose at an early point in civilization, when sexual impulses had to be controlled. They could not be entirely denied, but were made so unacceptable that they were rendered unconscious. That solved the immediate problem, but, Freud warned, these old instincts continue to rise to conscious attention from time to time, causing the sorts of problems he identified in his clinical studies and described as 'the return of the repressed'.[59]

Freud also focused on faeces and suggested that, more than any other single symbol, these represented what had been repressed. He pointed to the fact that young children are still fascinated by their faeces, and that most of us do not necessarily find the odour of our own offensive. Some of us continue to spend inordinately long periods of time in our toilets, keeping reading matter there for that purpose, and reserve our disgust for other people's products.

Confusion in this area is readily apparent in a number of psychic disorders, which could indeed be the result of strong pressures brought to bear when we first faced the problem of hygienic sewage disposal. Alain Corbin's study of eighteenth-century France suggests, however, that this problem had not been solved in any real way, even by that time.[33] It is still one that continues to plague communities in many urban areas. But if there is one thing that characterizes the human condition better than any other, it is that we are extraordinarily adaptable, even to bad odours.

In biological terms, what seems even more significant is that most smells are, and always have been, processed in unconscious parts of the brain: in the limbic system, which deals also with the organization and expression of reproductive behaviour and the emotions. And in this respect Freud was right. There are close ties between smell and sex, and between sex and emotion, so it is not surprising that sexual repression and olfactory repression should have become so intertwined. But what we need to focus on now is the direct and equally powerful connection between our sense of smell and our emotions.

The most obvious evidence lies in our fervent use of both incense and perfume, whose odours inspire us without any recourse to consciousness and the cortex. We take pleasure in them, both sacred and profane, because their nature is not only subliminal, and therefore beyond blame, but implicitly and inextricably sexual. It is hard to imagine anything more potent than secret sex that can also claim to have social sanction.

A good part of the influence exerted by perfume seems to lie beyond conscious control. All authorities hate that, for it threatens their programmes for us. The Song of Solomon is an evocative celebration of scented sexuality, composed at a time when ritual odours also played an important part in Jewish religious life. So it is not surprising to find the priesthood later complaining about the secular use of perfumes by 'painted Jezebels', who are accused of pagan worship and following the cult of Baal.

The wayward wife of Ahab, King of Israel, is not the only one to be tarred with this fragrant brush. In 188 BC, a general edict was issued forbidding Romans all but the most modest use of perfumes in social ceremonies. And when Mary Magdalene, now recognized as the patron saint of perfumers, anointed the feet of Jesus with spikenard, he was forced to defend her action by giving it a sacred significance. But the imputation remained, and still persists in puritan circles, that if a perfume is not intended specifically for religious use then it must be profane. As, of course, it is.

In 1770, the English Parliament passed an Act that protected men from the guiles of 'perfumed women' who might trick them into matrimony with the 'witchcraft' of scents that could manipulate their minds. This attitude carries all the way through to a paper published in 1913 in the *New York Medical Journal* on 'Connections of the sexual apparatus with the ear, nose and throat':

The use of perfumes from time immemorial has been a conscious or an unconscious attempt to stimulate lecherous thoughts . . . hence we hear of 'intoxicating' perfumes, well named, though few of those who use the phrase know the solid background upon which the expression rests.[39]

Everyone seems to recognize that perfumes do have a hidden agenda. Some object to this on purely social grounds. Socrates was concerned that the growing use of perfumes blurred the boundary between freemen and slaves. Muhammad bade his followers to wait until the next life, when all the faithful, regardless of their rank, would enjoy a paradise complete with 'houris made of musk'. But it was always impossible to stem a fragrant tide whose ingredients were so abundant and so accessible.

Take myrtle as an example. This little evergreen shrub is common in the Mediterranean area, but everywhere it grows it has come to be seen as something special. In myth, Aphrodite, she of the aphrodisiac, sought cover behind a myrtle bush when first she emerged naked from the sea. The Greeks called the plant *murto*, a name derived from the same root as perfume, because of the fragrance of the small white flowers and their volatile oil. Dioscorides, the author of *De materia medica*, the main pharmacology text for fifteen centuries, described the properties of myrtle oil as antiseptic, refreshing and aphrodisiac when taken in tea.

It is still widely used in the Middle East as a bath oil, and said to cleanse the skin, revive the spirits and keep insects at bay. The fragrance is fresh and clean and can be enjoyed by crushing the leaves underfoot, or by wearing a sprig in a garland as some Jewish brides still do. And it is the primary ingredient in 'angel lotion', a sixteenth-century deodorant that continues to be sold at country markets in Southern Italy.

What myrtle does, essentially, is to provide inexpensive, easy access to a fragrance that is both refreshing and stimulating. The top notes are cool and green, but beneath these, as in any popular perfume, there lies a base note with a hint of ancient ardours. These are the same fundamental chemicals that occur in henna, lime and chestnut, offering a hint of something faintly faecal, but definitely disturbing. Something that stirs up the limbic system, setting up a resonance about which the conscious mind can only wonder, adding even further to olfactory ambiguity. The nose knows things that we know nothing of.

The list of essential oils and their evocative odours is long and spans the alphabet from angelica to ylang-ylang, wafting through bay, caraway, dill, elemic, fennel, geranium, jasmine, linaloe, mustard, neroli, orris, patchouli, rosemary, sassafras, tansy, valerian and wintergreen.

It does these magical fragrances no favour to reduce them to esthers and aldehydes. But they do share the same peculiar chemical architecture, carrying ten atoms of carbon and sixteen atoms of hydrogen in every molecule. The rest is geometry, the result of nature shaking the raw materials into new patterns in its chemical kaleidoscope, with the result that one pelargonium smells of strawberry, while another variety smells of peppermint. Both are potent, but one is antidepressant, the other insect-repellent. Imagine what happens when you casually combine jasmine, carnation, gardenia and ginger in a single Hawaiian *lei*! The chances are, however, that this traditional floral recipe is not accidental at all, but the result of centuries of painstaking trial and error.

Animal odorants are mostly ketones, which are so closely related to plant aldehydes that they belong to the same group of organic compounds, making the journey from plant to animal matter as short as the shift of a single hydrogen bond. The difference between a plant oestrogen and an animal oestrogen is negligible. And musk can be found under the skin of a deer with tusk-like teeth, or in the seeds of several species of musk mallow and hibiscus. Chemically, the musks are hard to tell apart, but perfumers insist that the oils of animal origin have greater 'warmth and character'.

They may be right. Perfumery is still more of an art than a science, and it takes creative imagination to transform a number of olfactory nuances into the essential harmony of a fine fragrance with a personality of its own. The vocabulary of fragrance is filled with flexible words like freshness, refinement, delicacy, elegance, intimacy, opulence, sensuality, spirit. So Diva is opulent and intimate; Charlie, spicy and vigorous; Obsession, heavy and amberous.[198] I believe I can understand these descriptors. They have a sort of aromatic logic, but the most striking feature of 'smellspeak' is that it is wilfully, deliberately and, perhaps inevitably, ambiguous.

The perfumer Stephen Jellinek admits as much. 'Perfumes,' he says, 'serve as signals that carry multiple messages ... and it is a general property of signals that they may be misunderstood.'[86]

A scent, for instance, can say 'Notice me!', drawing attention to the wearer. But if it speaks too loudly, if the odour is too aggressive, it can backfire and label the wearer as someone trying too hard.

A perfume can use erogenic ingredients to say 'Love me!'. However, research suggests that this can sometimes frighten those who may no longer believe in witchcraft, but are uncomfortable anyway about feelings over which they seem to have little control. Most people fervently deny using perfumes for sexual attraction. When asked, they say it has more to do with their own confidence, but it is undeniable that having the power to attract someone else is very good for self-esteem.

An odour may suggest 'I am sophisticated!', manipulating a wearer's image by using the 'right' fragrance in a culturally approved way. The mere act of wearing a perfume marks the wearer as someone already a step ahead of animal nature. And wearing a scent that is recognizable as fashionable and expensive puts the wearer into a select in-group. But it also runs the risk of condemning that person as someone who lacks individuality, and is simply a dedicated follower of fashion.

All this theorizing, however, founders in the end on the fact that even those who claim to be able to recognize specific scents can very rarely do so. A large study of young women in Germany who were all regular perfume-users found that, when shown the package, most claimed familiarity with Opium, for instance, but just 20 per cent were able to identify that perfume in a blind test. Only 2 per cent could pick out Shalimar, and not one could recognize the classic Chanel No. 5, though more than half of those tested knew the name.[85]

The main function of advertising, marketing and image-making in the perfume business seems to be to send a message to the user, not to any third party who may smell the aroma in question.

The vast majority of women say that they use perfume simply because 'it feels good'. And maybe that's exactly right. It makes more sense than any signal function. All popular perfumes have enough of the right biological ingredients to be interesting at an unconscious level, so we have to assume a certain degree of pheromonal activity. But that is not all that is going on. There has been too much emphasis on the perfume itself, and too little thought given to the importance of the ceremony of putting the perfume on.

Ritual is important in our lives: it concentrates attention on the moment, on the things we actually do. And the perfume ceremony is very ritualized. It happens only on special occasions, and starts with a package filled with promise. The essence goes on slowly, drop by drop, dab by dab, applied deliberately to parts of the body as dictated by tradition, unconsciously honouring the points of pheromonal origin in a time-honoured sequence, pulling on the aroma like a costume, taking on the attributes of the image, becoming wicked and daring in tune with Obsession, romantic and passionate as Narcisse requires, and doing all this boldly in front of the mirror, in delicious disregard of religious convention, purely for the pleasure of it all.

Putting on perfume is a profane ritual. There is an element of sympathetic magic, too: you grow to fit the part for which the fragrance prepares you. The ceremony is also a rite of passage. It takes us out of the new artificial confines of a society with low olfactory tolerance. We have become so insecure about community, so intent on preserving our individual identity, that we habitually avoid most circumstances which force us to breathe the same air as strangers. We try to eliminate odours as a way of denying the hard truth that odour has no boundaries. Smell is continuous, and by deliberately putting on a ritual fragrance we bridge the gap between ourselves and others. We make the transition from 'me' to 'us'.

Smell is life-affirming and almost self-fulfilling. It works in both directions. Perfumes, with the help of Jacobson's Organ, have an unconscious impact that changes our feelings, our body chemistry and our body odour.

Alan Hirsch at the Smell and Taste Treatment and Research Foundation in Chicago has found that a floral scent helps volunteers to perform puzzle-solving tasks 17 per cent more quickly; and that aroma engineering in a Las Vegas casino can increase customer optimism, and therefore willingness to gamble, by as much as 53 percent.[198]

International Flavors and Fragrances of New Jersey, now the world's largest fragrance company, offers its clients 'Muzak for the nose', flooding the workspace with carefully selected smells. The results are surprising: the Good Housekeeping Institute reports that proofreaders are more accurate when reading to the accompaniment of peppermint or lavender odours. And in Japan, Takasago International has discovered that a whiff of lemon wakes the staff up first thing in the morning, the comforting scent of roses prepares them for lunch, and jasmine later in the day is uplifting for tired keyboard operators. And it is interesting that, everywhere, these beneficial effects are still felt even when the scent is too faint to come to obvious attention.[85]

So, at the Monell Chemical Senses Center in Philadelphia, Gisela Epple is studying the effects of specific odours on children under stress. At the University of Minnesota, Mark Snyder is working on problems that arise between couples whose scents have begun to bother each other. And at Bowling Green State University in Ohio, Peter Badia is experimenting with the effect of smells on sleep and dreams. The results of studies like these have already helped the Sloane–Kettering Cancer Center in New York to relax patients with the fragrance of vanilla. And at Duke University, a hint of menthol in the changing room seems to be paying dividends in preparing athletes for big events.[212]

All these modern applications of the ancient traditions of aromatherapy are interesting. They are predictable spin-offs from a fragrance industry that produces over five thousand separate products worth $5 billion every year. This is very big business, but the golden grail of the industry is the fragrant version of Viagra: something that can be sold over the counter to produce great sex with a single sniff. So far, no luck, but the scramble is on, and one of the best bets could be a small venture company called the Erox Corporation in Menlo Park, California.

Their story starts in the early 1960s at the University of Utah, where biochemist David Berliner was working on the chemical components of the human skin. He had several flasks of skin extracts on his desk and, for the first time in months, the mood in the laboratory was so relaxed that workers got together for a game of bridge over lunch. Everyone felt good about this, and the games went on until Berliner reached the end of that phase of his work.

'Then,' he recalls, 'I put the flasks back in the freezer, and bridge stopped automatically. No more bridge.' Several months later, he brought the flasks out again, camaraderie revived, and the card games started up once more. Berliner noticed the connection, but did not pursue it as he became involved instead in work that resulted in the development of the skin patch for delivering drugs to treat nausea, heart disease and nicotine withdrawal.[214]

Twenty-five years, several companies and many millions later, Berliner decided to thaw out the original flasks and see if he could find out what it was they contained that might have been responsible for such unusual bonhomie. In 1989, his first assay of the material isolated twenty natural substances, some of which seemed to leave volunteers less nervous and more confident after a few casual sniffs. He thought they might be human pheromones, but more careful research was needed.

In 1991, Berliner published a paper reviewing the nature of human skin, examining its chemistry and pointing out that each of us loses forty million surface cells every day. These, he suggested, would provide a perfect pheromone delivery system, one ideal for exactly the sorts of non-volatile, non-odorous compounds that could best be received by Jacobson's Organ.[12]

So Berliner joined with David Moran, Larry Stensaas and others at the University of Utah in an intensive search for the organ in human noses. Moran was already looking at Jacobson's Organ, finding it present in almost every human subject, and wondering how to prove that it was functional. Luis Monti-Bloch designed an apparatus to deliver selected chemicals to the tiny pits of Jacobson's Organ and detect any electrical activity that followed. He called it the electrovomeronasometer, and found that it recorded nothing for any smelly substances. Then Berliner gave him some samples from the famous flasks, and suddenly the machinery burst into life.

It was clear that neurons in Jacobson's Organ were sparking in response to these samples, and as a direct result of this activity there were changes in heart rate, respiration, pupil size and skin temperature in the owner of the nose. Monti-Bloch published the results in 1994 with the casual conclusion that 'these indicate that the adult human vomeronasal organ is a functional chemosensory organ'.[128] In other words, the organ that textbooks have ignored for most of the century not only exists, but responds to selected biochemicals.

At that point, Berliner was still being very cagey about the real nature of his samples. But as soon as he had secured patents on them all, he and Monti-Bloch joined in the formation of the Pherin

Corporation for further research. And in 1996 they published the keystone paper which demonstrates that substances they called steroidal vomeropherins from human skin can and do attach to receptors in Jacobson's Organ, which sends messages about them to the limbic areas of the brain.[13]

Berliner's substances elicit responses in the autonomic nervous system. These differ from person to person, and show sex differences of the sort one would expect from pheromones. They look like and work like pheromones, and should be enough to convince most neuroscientists of the reality and sensitivity of Jacobson's Organ. All that is holding up a scientific consensus seems to be the fact that Berliner has also spawned a pharmaceutical company called Erox to produce two commercial fragrances.

These are called Realm Men and Realm Women, and are being advertised as the only fragrances in the world to contain human pheromones to be duplicated in the laboratory instead of being borrowed from the glands of musk deer or civet cat, or extracted from plant resins. Erox is being very careful not to fall foul of the Food and Drug Administration's requirement that any product calling itself an aphrodisiac can be sold only as a prescription drug. The pair of matching perfumes are offered only as products designed to boost the positive feelings of male or female users.

Each one is said to be detectable only by the sex in question. Realm Men contains a steroid from women's skin and is worn by men, for their own self-esteem; Realm Women contains a steroid from male skin and is intended simply to boost the confidence of the women who wear it. In both cases the scent presumably provides each sex with the pheromonal illusion of having recently been involved in intimacy with the other. Taken together with the user's own body odour, that appears to be a very attractive combination. First reports suggested that trade is brisk. The problem for Berliner is that science is notoriously uncomfortable with the idea of any of its practitioners becoming venture capitalists and making huge sums of money out of their discoveries.

And that, for the moment, is where the matter rests. We appear to have a working Jacobson's Organ and a fully functional second olfactory system. But the news is very slow to spread. That surprises me, because the implications are staggering. If it is true, this is a discovery that could change our lives.

PART THREE
THE MOST HUMAN THING

Every time an olfactory issue is raised, it is worth reminding ourselves that hearing and vision are, in evolutionary terms, relatively new senses. They are the playthings of our new, big brains and they monopolize our intellectual and aesthetic attention. There is no equivalent role for smell in our mass culture: scratch-and-sniff is no threat to *son et lumière*. But this neglect is not a result of olfactory insufficiency. It seems to me to be a reflection instead of the primary nature of smell. Smell is too important, too fundamental to our basic well-being, to be toyed with.

The truth is that odour is still extraordinarily powerful. And this power has nothing to do with repression or forgetting – it is the product of an evolutionary trend towards a new way of dealing with smells, a way that is adaptive and creative, and unique to our species. It has to do with memory.

Neurological studies suggest that, unlike those in other animals, very few of our olfactory circuits are 'hard-wired'. We are born with a tendency to respond to the smell of our mothers, and we grow into a world of new odour preferences at the time of puberty. Very little else is fixed. The rest we have to learn, largely by trial and error and by social example. We acquire our odour memory from experience. Where we keep it remains a mystery, but it appears that far more of the brain participates in odour perception and retention than was previously thought. Electrical mapping of the human brain is still in its infancy, but it has shown that smells seem to touch both cerebral hemispheres: exciting emotional activity on the right side at first, leading later to intellectual activity, such as trying to remember the name of a particular perfume, on the left.

None of this research, however, comes close to understanding what might be called the 'Madeleine effect'. Marcel Proust is best remembered for his *Remembrance of Things Past*, a huge and influential novel of seven volumes and many memories, written in the unventilated bedroom in Paris where he spent most of the final ten years of his life. In it, he discusses his preoccupation with the 'lost landscapes' of remembrance which lie beyond voluntary recall:

When, from a long-distant past nothing subsists, after the people are dead, after the things are broken and scattered; taste and smell alone, more fragile but more enduring, more insubstantial, more persistent, more faithful, remain poised a long time, like souls, remembering, waiting, hoping, amid the ruins of all the rest; and bear unflinchingly, in the tiny and almost impalpable drop of their essence, the vast structure of recollection.[152]

What sparks this observation is an extraordinary *mémoire involontaire*, a chance savouring of the taste and smell of a madeleine, a sponge cake, dunked in *tilleul*, or linden tea. This little citrus-smelling madeleine brings back all the rich, sensual background of the village of Combray, where the narrator spent his childhood. It triggers memories more detailed and evocative than vocabulary can ever reach, conjuring up multi-sensory, whole-flavoured incidents, heavy with all the erotic sensations of the natural world.

This 'Proustian moment' resonates now through literature, helping other writers to prise open the locked doors of their own original experiences, exhaling recollections and preparing new ground for cultural and natural history, helping us all to put up what Proust called 'a nightlight in the bedroom of memory'.

Odours are indeed guardians and gatekeepers of the past. At Warwick University in England, volunteers were put under stress by being asked to complete an intelligence test in an impossibly short time. Half of them went through this experience in an atmosphere with a low-intensity neutral fragrance. Several days later, *all* the volunteers went through a similar test with the same faint odour in the air. In this second session, those who had not smelled the fragrance before did better at the now more familiar test, and found it far less stressful, while those who were being exposed again to the odour showed even greater anxiety, and returned even poorer scores.[100]

The subjects were being unconsciously conditioned by an odour none of them noticed or could later remember smelling. Even unfamiliar and unremarkable odours, it seems, can be associated with stressful situations and become signal smells, reproducing or reinforcing that stress on a later occasion. So, if the smell had been a biologically significant one, perhaps one released by a vaginal secretion, it would forever become sexually significant and exciting to a young man whose first exposure to it had been in bed with a woman. If the experience had been pleasant, it would have become an aphrodisiac. If the experience had been a bad one, the same, chemically identical substance could just as easily have become repugnant – even though, in itself, it need not have had either attractant or repellent qualities.

Our memory of such odours is apparently simple, direct, unconscious, even 'primitive', and very resistant to decay or later interference. This is certainly what makes many long-term memories so hard to articulate: they have a 'tip-of-the-nose' quality. We recognize them, hazily, but find them very hard to label or identify except as odours associated with a particular experience. They and their effects are acquired under circumstances below the level of verbal awareness. They go directly into the limbic system, which means that many of them are almost certainly perceived and transmitted to the brain by Jacobson's Organ.

Steven van Toller at Warwick University has found that in some of his olfactory experiments, subjects show clear electrical activity on the skin and in the brain, while denying that they have received the odour that provoked the response. In some cases this seems to be because there was no conscious awareness of the smell; in others, the denial is of a smell for which the subject has no verbal label or convenient category. So, despite the fact that the brain registers a smell, and may even recall it under hypnosis, it 'ignores' the odour simply because it is 'unspeakable'.[199] You can smell it, but if it doesn't make conscious sense, it is nonsense, it hasn't happened. Selective anosmia is one of the trademarks of the limbic system.

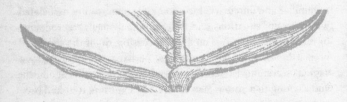

John Kinge, another member of the productive Olfaction Research Group at Warwick University, is concerned with the elusive connection between smell and long-term memory. He works with the elderly in a programme he calls 'reminiscence therapy', using a nostalgia aroma pack designed to evoke smells of the early 1940s. Trigger odours are held in stoppered vials containing pads soaked in scents identified simply as 'Old Teapot', 'Washday' or 'Air Raid Shelter', each of which is capable of conjuring up long-forgotten information. After a sniff of an old-fashioned disinfectant in the sample labelled 'Field Hospital', one veteran found to his astonishment that he could still rattle off his wartime rifle number.

Group sessions with the aroma pack are even more productive. The unique odour of a teawagon that once belonged to a wartime women's auxiliary service catalysed a recollection-fest among several women in their nineties. 'It's amazing,' said one. 'With a single sniff you start to think again about all sorts of things. Memories come pouring out of the back of your mind. It was such a treat.' The charred smell of bombed-out buildings in the vial identified baldly as 'Blitz' is another favourite with Kinge's patients. 'As the population ages,' he says, 'we are going to find a lot more old people with memory problems. Smells touch the memories that other senses cannot reach. I know of no more effective way of re-orienting people.'[158]

Providing the same service fifty years from now may not be so simple. For those growing up in fragrance-free zones or in homes where most food is microwaved instead of oven-baked and there is no lawn to mow, there seem to be far fewer characteristic odours to which memories can become attached. It is possible that this deprivation may even deny today's young people their reminiscences, but it would be foolish to underestimate the creative capacity of olfactory memory. Who knows what the smell of diesel fumes or french fries might mean to those looking back from the year 2050?

The peculiar strength of smell is that it is the only sense that has direct contact with the brain. So it is not surprising that memories of stimuli received there, direct from their sources, without processing or model-making or image-bending by the conscious brain, are very different from the memories linked with sight or sound. A fragrance is not organized spatially or modulated temporally: it is an experience of the moment, with none of the usual textures of time and space. And being free-form it is also very difficult to encode or file away in the usual places.[23]

The problem here is very much like the 'observer effect' in quantum mechanics, which suggests that it is impossible to be objective: the mere act of observing changes that which is being observed. Just by taking part in an experience, you inevitably become involved. So when you smell a fragrance, you change it and it changes you. The experience may be hard to describe, but it will also be impossible to forget because it is imprinted somewhere deep in the brain in a form which gives it extraordinary clarity. Odour-linked memories in humans come with all their flavours intact, even after many years. We may fall in love 'at first sight', but it is the familiar smell of a loved one that sends our blood pressure rising.

Smell is an emotional sense, rather than an intellectual one. It is more right brain than left, more intuitive than logical, and more open to synesthetic combinations with other senses. And when it involves Jacobson's Organ, it is more likely to be unconsciously than consciously perceived. All of which appears to make odour-linked memories impossible to forget.

They constitute a vital, primitive protection system, making it possible for us to learn in just one short trial that something is poisonous, and must forever be avoided; helping us to recognize kin, find food, track prey and keep tabs on the reproductive status of those around us. The system is highly adaptive, with a strong ancestral purpose that gives it great evolutionary significance.

But in the end, what matters most about the madeleine effect is that it parlays a primitive pattern of conditioning into a formidable creative tool. It gives us access to the past with a clarity, and the kind of total recall, that no other memory system can match. It has turned the nose, formerly just the body's advance guard, into an organ of wisdom; and smell, once dismissed as good only for letting one know when the toast is burning, into the sense of imagination.

Looked at in this way, smell begins to assume a more dominant position in the ranking of our senses. We may have access to less olfactory information than dogs or horses enjoy, but our big brains allow us to make far more of the news we do receive.

Alliaceos

Before Linnaeus, botany and medicine came down through the ages hand in hand. Then both arts became sciences and went their separate ways. Botanists ignored the medicinal properties of plants, and the medical texts refused to consider herbal lore. But a few plants were so potent that they took root on both sides of the modern divide.

One of these was garlic, the source of the antibiotic allium and the fungicide allicin, and in addition perhaps the most mystically active and attractive bulb in the business. Garlic is a lily, with up to twenty edible cloves or bulblets packed with magical oils so penetrating that when a clove is rubbed on the soles of the feet, its odours are exhaled by the lungs.

Theophrastus tells of Greeks leaving garlic at the crossroads as a supper for Hecate, three-headed goddess of the underworld. Homer has Odysseus take garlic to avoid being turned into a pig by Circe, the daughter of the Sun. But anyone with garlic on their breath was forbidden to enter any of the temples of Cybele, the mistress of wild nature.

Today, the air of the entire Mediterranean is perfumed by people who eat garlic in food, drink it in wine, use it for asthma, chronic bronchitis and dropsy, even rubbing it on their skin to prevent suffering from the debilitating effects of the ill winds sirocco and simoom.

On these final pages, 'poor man's treacle' appears at the pinnacle of the Lynnean pyramid of odour-types, because though it may frighten crows and scare moles up out of the ground, garlic is everyman's easiest and most interesting route to good health and fragrant well-being.

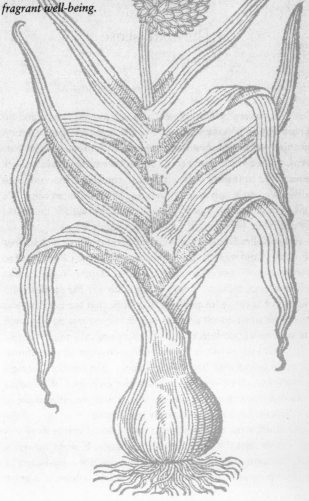

7

The sixth sense

Our grasp of reality owes almost as much to politics as it does to physiology. We *make* sense of life. We break the news up into bite-size pieces and then reassemble them into meaningful patterns. We see what we expect to see, not what the lens in our eye detects. We hear what we want to hear, tuning out inappropriate noise. We smell only odours that are new and interesting. We make conscious and unconscious choices, voting for the world we need. But sometimes the choices are made for us.

Helen Keller was stricken by a mysterious illness that left her blind, deaf and mute before her second birthday. Condemned to a sensory limbo not unlike that of Kaspar Hauser, at six she was still untamed, kicking, biting, eating with her hands. But from the age of seven, with the help of a dedicated teacher, she was able to let smell, touch and taste guide her excursions into what she later called 'the borderland of experience'. She smelt 'the tide of scents which swells, subsides, rises again wave on wave, filling the wide world with invisible sweetness'. She felt the nostalgia induced by 'scents that start awake the memories of summers gone and ripening grain fields far away'. And she pitied those of us who can see, but are effectively smell-blind.

She could sense the coming of a storm, hours before there was any visible sign, by 'a throb of expectancy, a slight quiver, a concentration in my nostrils'. She could describe a landscape by the grouping of its odours, putting a hayfield, a barn or a grove of pines in their proper places. She could distinguish a carpenter, an ironmonger, an artist or a chemist by the fragrance of their trades, and 'When a person passes quickly from one place to another, I get a scent impression of where he has been – the kitchen, the garden, or the sick-room.'

She was at her most responsive in meetings with other people. 'If many years should elapse before I saw an intimate friend again, I think I should recognize his odor instantly in the heart of Africa.' She found her sense of smell as difficult to put into words as the rest of us, admitting that:

Some people have a vague, unsubstantial odor that floats about, mocking every effort to identify it. It is the will-o'-the-wisp of my olfactory experience and I seldom find such a one lively or entertaining. On the other hand, one who has a pungent odor often possesses great vitality, energy, and vigor of mind.[95]

Helen Keller even had the courage to contemplate the loss of her 'potent wizard', saying: 'I know that if there were no odors for me, I should still possess a considerable part of the world. Novelties and surprises would abound, adventures would thicken in the dark.' The triumph of her spirit inspired millions and still makes an important point, one lost on physiologists who insist that there is no evidence to show that blind people are able to smell any better than sighted ones.

That conclusion is possible only if you measure nothing but the threshold for particular odours isolated in a laboratory situation. Then our senses all hover around an average level for our species, with no real difference between amateur sniffers and the trained noses of professional perfumers. But what such assessments ignore is the context. Take wine tasters out of the laboratory and put them instead into a cellar where they can look at the colour of the vintage, see how it swirls in a glass, and savour the aftertaste on their tongues. Then the gap opens wide as they show beyond doubt that they can do things impossible for any untrained person. They can tell the 'nose' of a burgundy from a bordeaux, placing it not just on a particular estate, but even in its year of origin.

We very seldom, if ever, experience any sensation in isolation. Our senses feed on one another, shifting, sorting, reinforcing, blending as they go. Smell may have been our first long-range sensory experience, but now all the domains overlap, sharing common feeling-tones and recreating a primordial unity of the senses. And strangely, nothing is lost in the mix. The separate senses are not so much diluted as reinvigorated, giving someone like Helen Keller the chance to 'see' as well as any sighted person.

This is not news to poets, musicians and artists, who already inhabit synesthetic worlds. Charles Baudelaire described 'Perfumes fresh as children's flesh, sweet as oboes, green as prairies.' Guy de Maupassant admitted, 'I truly no longer know if I breathed music, or if I heard perfumes, or if I slept among the stars.' Percy Shelley celebrated 'music so delicate, soft, and intense, it was felt like an odour within the sense,' and likened the song of a nightingale to 'field smells known in infancy'.[156]

Nicholas Rimsky-Korsakov and Alexander Skriabin agreed that E major was blue, though they differed on its exact hue. Walt Disney squeezed a whole palette out of Bach and Mussorgsky for *Fantasia*. The Symbolists, Stéphane Mallarmé and Tristan Corbière among them, anticipated by a full century the psychedelic synesthetic enthusiasms of the 1960s. But it was A. E. Housman who brought everyone back to earth with his frank admission that poetry was, in fact, not an art at all, but far more visceral – more like 'a secretion'.

All this poetic licence is fun, but perhaps the best evidence we have to suggest that there is a biological basis for synesthesia comes from linguistics. Joseph Williams at the University of Chicago has analysed adjectives in English which refer to sensory experience, and discovered what he describes as a 'semantic law', something that helps us to understand how sense and meaning can change over time.[213]

Good, hard laws in linguistics are very rare, but going back twelve centuries, Williams found that there is a directional flow in the transfer of adjectives from any one of the five senses to any other. Among chemical senses, the flow goes from touch to taste to smell, but never in the other direction. So a 'dry' touch effortlessly slides into a 'dry' taste, and even to a 'dry' smell. But neither adjective slides back down the scale to touch. Tasty words can become odorous, but never tactile. Almost every adjective from touch and taste moves up to attach itself finally to smell, but no known primary olfactory word in English has ever shifted to any other sense.

Smell exists at the end of the semantic line, just as it lies at the head of a biological flow that goes from touch and taste to the more sophisticated receptors involved in smell. The newer senses of sight and sound seem to show a separate and parallel pattern of semantic development, receiving adjectives directly from, and only from, the primitive sense of touch – as in 'warm' colours or 'soft' sounds. But second-level transfers beyond sight and sound take place only between each other, producing mutual overlaps such as 'bright' sounds and 'strident' colours.

Williams has found this same clear pattern in Latin, Greek, High German and Japanese, with word shifts in every case mirroring the sequence in which the senses come into action in our individual lives. First touch, then taste, then smell, and finally sight and sound. First the hagfish bumps into things in the mud, then it makes the taste test, then it investigates more distant smells. First the human infant turns towards the touch of a nipple, then it tastes milk, then it recognizes its mother by her smell. Then fish, child and adjective all go on to the new excitements of sight and sound.

It is fascinating that language evolves in much the same way as any other living thing. And significant, too, that it finds advantage in being upwardly mobile, mixing only with more experienced words associated with more highly developed senses. In this sensual hierarchy, smell is in an odd position. It makes direct connection with the brain and has no need to piggyback sight or sound, but can benefit from synesthetic vigour and, lacking any other alternative, it appears to favour joining forces with our other olfactory system. Jacobson's Organ already alternates with the olfactory epithelium in bloodhounds, producing new and more potent sensory skills. Could it do the same for us?

We know that our noses can be educated. William Cain at Yale University chose eighty odorous substances with which every American ought to be familiar, and found that his students, on average, could identify only thirty-six of them by name. Most of the rest they knew, or thought they knew, but couldn't name. But all it took to bring the average score in later tests up to 94 per cent was a little gentle prompting and some positive feedback to wake up dormant abilities. All of us, it seems, have a talent for odour identification, and can demonstrate this with the right kind of encouragement.[22]

In Japan they play a game, called *kodo*, in which participants try to guess the names of up to two and a half thousand different scents. In even more refined versions of the game, one kind of incense is burned and the participants are required to find whichever other incense will combine most pleasantly with the first. And at the highest and most aesthetic level of competition, participants try to create an odour which would be most appropriate as an accompaniment to a particular literary work, perhaps one of the lesser-known poems of Kokinshu. *Kodo* gives a good idea of how little use most of us make of our everyday olfactory potential. Rachel Carson, enjoying the power of the sense of smell to bring back memories, said, 'It is a pity that we use it so little.'[24] But perhaps it is not so much a case of disuse as one of lazy recall.

There is obvious advantage in being part of a culture, like that of Japan, which finds new ways of celebrating a neglected sense. There is merit in the idea of bringing smell to conscious attention by introducing *kodo* or other odour-training exercises into our schools. And it helps, of course, to be compelled, like Helen Keller, to make the best of what we have. But in the end it may be unnecessary to go out looking for ways to improve our olfactory performance, simply because everything we need is already there in place, right under our noses.

We have Jacobson's Organ. It is present in almost every human nose so far examined. And it appears to be functional: a miniprobe inserted into the appropriate pits on either side of the nasal septum registers a measurable electrical potential in the presence of steroidal pheromones and some other molecules. Particularly those of human origin, which tend to be less volatile and therefore more likely to impact on this organ near the nostril, than on the olfactory epithelium at the back of the nasal cavity.

And all that remains is to imagine what it might mean to own and use a hitherto unsuspected sense organ! My guess is that our lives will never be quite the same again. Just writing and thinking about this has opened new sensory vistas for me. But everything that follows here is pure speculation on my behalf . . .

Just imagine being able to smell direction, and literally 'follow your nose'. Vertebrates were the first to have twin nostrils opening onto a pair of nasal cavities separated by a septum into two olfactory streams, providing 'stereo smell'. This is a very useful talent. With it, we can track down a smell source, find food, locate mates and escape danger.

Directional smelling has been selected for in fish, reptiles, birds and many mammals, and seems even to have survived our transition to a less nomadic way of life. Tests with odours applied to only one nostril at a time show that we can distinguish very well between two streams, isolating a scent to either the left or the right side with a failure rate as low as two in a thousand. And even when the scent in question reaches both nostrils, there is a slight delay in recognition time and a perceptible difference in the intensity of the stimulus. We may not be conscious of doing so, but we can tell direction and pinpoint a smell source.[103]

Imagine, then, an odour with the kind of molecular structure and hormonal quality which also reaches Jacobson's Organ and the alternative olfactory system. On biochemical grounds alone, this will be perceived more slowly than a volatile odour, and travel anyway to the limbic system before being brought to cortical attention. I have already mentioned the survival value of such a delay in providing life- and face-saving 'second thoughts', but it is just as likely that this differential will also provide information that can support or dispute directional conclusions arrived at by the nose alone.

We know also that it is rare for both our nostrils to be working at full capacity at the same time. Each appears to dominate for a while, changing over roughly every three hours, contributing to a biorhythm that has a direct effect on mood and behaviour. Normally this balances out, but when one nostril is obstructed by injury or infection, the other suffers from fatigue and fails to carry even its own share of the olfactory load with the same sensitivity. That compounds the problem, contributing to the disorientation we experience during colds or influenza.

But Jacobson's Organ doesn't depend on a continuous stream of air, and if it continues to function on either side of the nose it could pick up some of the slack, providing information ultimately to both sides of the brain. It could also play a cyclical role of its own, alternating from left to right as necessary, standing in for the nasal epithelium under certain circumstances, or even on a regular basis, providing another source of rhythm for bodies that thrive on any information that can help to maintain homeostasis.

Think about it. The discovery that plants communicate with each other by means of airborne hormones very similar to our own gives pause for thought. And the almost simultaneous rediscovery of Jacobson's Organ as a receptor system for such hormones begs for connection. Why not? It would help explain a lot of strange things that have been going on.

For instance, it is over a hundred million years since Madagascar was isolated from Africa and drifted off into the Indian Ocean, carrying a cargo of animals and plants that no longer exist anywhere else, or have evolved into an extraordinary number of endemic forms. Humans are very new there, having arrived only in the last two thousand years to confront an exotic flora of almost fifteen thousand species of flowering plants, 90 per cent of which were totally unfamiliar to the immigrants from Africa and Asia. And yet these newcomers have managed to compile an impressive pharmacopoeia of useful herbal remedies, thousands of which are on sale in every country market. There just hasn't been time to test every strange plant on the island and decide which part, of which species, picked in which season, and prepared in which way, would be appropriate for which human conditions.

Such decisions have been arrived at in Asia and Africa by painstaking experiments over thousands, perhaps even millions of years, drawing on long traditions of association with local floras. But permutations of all the possible variables run into the billions, making it impossible, in fewer than eighty generations, for newcomers to Madagascar to have arrived at their decisions by trial and error alone. They must have had help.

On every visit to Madagascar I have tried to talk to local healers, or *ombiasy*, and to learn something of their techniques. But when I ask, for instance, how they know that an extract from the leaves of a local flowering plant, picked in the spring, is good for a condition they describe as 'milky blood', I always get the same answer. 'Oh, it's easy,' they say, 'we ask the plants.' Absurd, of course. But that is exactly what they do. A healer with a problem goes out into the forest, thinking about the patient, and tries to sniff out a solution by wandering, with an open mind, until something stops him, until a particular plant catches his attention, and gives a whole new slant to the idea of an elective procedure, by offering itself as the remedy.

I was sceptical about the notion until I made two discoveries. The first was that 'milky blood' is, in fact, an accurate description of one of the symptoms of a condition we know as leukaemia. This form of cancer spreads from the bone marrow and floods the vascular system with immature white blood cells to such a degree that red cells are overwhelmed and blood does indeed begin to look a little creamy. And the second discovery was that a Swiss pharmaceutical company is having some success in the treatment of childhood leukaemia with an extract from a plant commonly known as the Madagascar periwinkle.

Coincidence? Perhaps. But I am beginning to suspect that plants are probably signalling more than just alarm, that there may be a far more democratic exchange of information going on in every ecosystem than we ever imagined, and that a lot of this information is olfactory. Perhaps some plants produce hormones, not as a warning to each other but as an invitation to an animal that might help in the plant's distribution. Most often this sort of signal is a sex attractant, aimed at a potential pollinator, but that need not always be the case. Nor is the reward always nectar: it could even be a medication, advertised by a smell detectable to a passing Jacobson's Organ. How else do domesticated dogs, with no direct experience of the wild, know which plants to eat when they have a stomach ache?

Maybe we have a similar talent, a licence to practise herbal medicine. When we need help, perhaps all we have to do to access the local Odornet is to exercise a little humility, and go out and ask. And if my experience in Madagascar means anything, it could be useful also to approach the plant world as *ombiasy* do – not just with an open mind, but with the upper lip slightly flared in a way that opens up the duct to Jacobson's Organ.

Vision dominates our lives, our language and our minds. We say 'I see' when what we mean is 'I understand'. We say 'Look' when we mean 'Listen'. And even in this age of computer simulation and special effects, we continue to have a touching faith in eyewitness testimony. It might help to redress this imbalance if we gave a little more credit to our noses.

Why, for instance, do we continue to shake hands as part of the greeting ceremony? Close friends, families and lovers all hug or embrace in demonstrations of their personal bonds. They get well within each other's odour envelope and enjoy the familiar smells. But the handshake is different. We shake hands even with unwelcome guests – in fact we go out of our way to shake hands with strangers and those we may hate or later come to fear. Princess Diana made news by becoming the first female member of the British royal family to shake hands without the customary white gloves, and with victims of Aids.

I suggest that the reason we attach so much ceremonial import-ance to the ritual handshake, and recognize the gloved version as a poor, even insulting, substitute, is that it involves skin-to-skin contact and a direct and important exchange of pheromonal material. Where smell is concerned, the gloves have to come off. Most other parts of the body are camouflaged by fragrant shields which conceal the real person, but very few people perfume the palms of their hands. And after all, after contact with someone 'distasteful' we can't wait to rush off and, like Pontius Pilate, wash our hands.

Take a close look at the next state occasion or diplomatic overture which brings together two men who may soon be at each other's throats, and watch what happens immediately after the official handshake. If the cause is hopeless and the outcome already decided, each will indulge in a surreptitious wipe of his palm on a trouser or convenient chair arm. But if the affair is not yet decided and there is jostling for position, one or both men will, immediately after contact, make an unconscious gesture that brings the ceremonial palm up close to his nose. He may adjust his spectacles, scratch a cheek, swat away an imaginary insect, or simply adopt a thoughtful posture with his hand on his chin – whatever it takes to get the crucial scent closer to the nose for analysis.

Part of this unconscious 'sniffing out' procedure certainly involves Jacobson's Organ. We need to know everything we possibly can about strangers in our lives who may turn out to be dangerous enemies. Knowledge is power, and I believe it is no accident that we are prevented, by the same traditional white cotton gloves, from making similarly personal assessments of their wives and consorts. This, too, is part of the process of concealing the reproductive condition of women from outsiders. And it helps to explain why Diana so scandalized the establishment, and why she was so revered by those with whom she came into direct pheromonal contact. She touched people.

Speaking of such contact, we can't avoid the fact that Jacobson's Organ specializes in reproductive information. In most other mammals, it traffics in news about who is ovulating, who is looking for a mate and even who is the real father of whose young. There is no reason to assume that all such information is not potentially available also to us. Removing the organ in mice, hamsters and pigs has proved to be as drastic to them as castration: it destroys their sex lives. Ignoring it or damaging it could have equally disastrous effects on ours.

The work that led biochemist David Berliner to manufacture

fragrances based on the steroids found in human skin shows that we already produce several natural hormones that can activate Jacobson's Organ in members of the opposite sex. This suggests that our sexual behaviour can be modified by the secretions of another individual, by someone with 'the right chemistry'.

Clive Jennings-White, one of Berliner's colleagues at the University of Utah, has identified two of these substances as androstadienone and estratetraenol, the first odorous, and the second completely devoid of smell.[88] The odourless steroid, taken from women's skin, produces a strong response in male Jacobson's Organs. The smelly steroid, taken from men's skin, produces a similarly strong response in female noses. So both pheromones are gender specific: each is intended for the other sex and works unconsciously on parts of the other person's brain. But both sexes are also able to smell an olfactory marker on the male pheromone, which happens to be the active ingredient in Erox's Realm Women perfume.

This disparity is interesting. It means, for a start, that the female pheromone produces its effect totally unconsciously: there is no marker scent to identify its presence or its source, maintaining a woman's anonymity and providing no clue to her whereabouts. The male pheromone, however, is not only a sexual signal, but also an active advertisement, a proclamation from the male that he is the sender, and that the road to her leads only through him. It is a clear territorial signal for the benefit of rival males who cannot detect the sexual component in his transmission, but are left in no doubt about the overt threat in the 'top notes' of his hormonal performance.

That is as far as current research goes, but even these initial facts give rise to some entertaining possibilities. H. G. Wells, by all accounts, was unusually attractive to women. His colleague Somerset Maugham wondered why, and asked several of Wells's admirers what his secret was. They all said the same thing – it was the way Wells smelled. The great writer's pheromones were probably no different from any other man's, but slight variations on top of the usual chemistry give everyone a unique odour, and the grace notes in Wells's presentation could have come from a number of sources. The active ingredient was probably one of the usual androgens, which women have learned to associate with men, and which remind them of the base notes in their own musky perfumes. But I would be prepared to bet that the clincher was complex, perhaps a typically Wellsian vocal overlay to the fragrance.

In our species, Jacobson's Organ is seldom enough on its own. We are more complex, more unpredictable, but secret scents remain a very important part of the whole package.

Without Jacobson's Organ, we are likely to be severely handicapped, even subject to developmental difficulties. The organ is clearly visible in the human foetus and it is demonstrably present in the nose of all newborn babies. And although it may be masked by the later growth of other features, it persists in almost all adults.

Jose Garcia-Velasco at the University of Mexico has made the largest study of the incidence of Jacobson's Organ. He looked at 1000 patients of both sexes seeking plastic surgery. Using only a nasal speculum and the light from a headlamp, he immediately found the tell-tale pits on both sides of the nasal septum in 808 of these subjects. Looking again at the remaining 192, he found that 125 had injuries or deformations of the nasal septum. They had come to the Medical School for reconstructive surgery, and once this had been completed, straightening out distorted septa, the pits of Jacobson's Organ became clearly visible in all but 23 patients. The Organ of Jacobson, Garcia-Velasco concluded, 'is a normal, distinct structure of the human nose, and is present in practically all subjects studied'.[60]

We have an unfortunate tendency to underestimate and devalue parts of the body for which we have not yet discovered a useful purpose. The appendix, the tonsils and the adenoids have all suffered from this rush to get rid of everything surplus to requirements, and each has been the focus of fashionable and often unnecessary surgery. But we become, if not wiser, at least more circumspect, leading one physiologist to conclude: 'Humans have a set of organs which are traditionally described as non-functional, but which, if seen in any other mammal, would be recognised as part of a pheromone system.'[30]

Things which seem obvious in moths become contentious in humans. This is largely because we are reluctant to admit that, in many respects, we are like other animals and often act without due deliberation – sometimes even producing a detectable odour. So we downplay our tendency to sweat and our capacity to respond to the body odour of others, and pretend that we are not like them. Very often, of course, we are. Our children grow up inside our odour envelopes and learn, with the help of Jacobson's Organ, to identify their mothers by smell alone. A failure to do so can be harmful. We already know that bottle-fed babies are less responsive than those who have been breast-fed. Social skills begin to be learned and practised from the moment of birth.

We learn a lot from our parents, often in a tacit fashion, just by having them around. And if the mere presence of a man in the house can entrain the hormones of the female family members, bringing girls to puberty at an earlier age, it doesn't seem outrageous to suggest that children growing up in a close family are going to be physiologically and psychologically different from others deprived of such biochemical contact. We have only just begun to explore the true nature of father–daughter and mother–son relationships and the possibilities inherent in the interface at Jacobson's Organ: our window on the world of pheromonal communication.

I suspect that this is where the tensions between step-parents and step-children, foster parents and their charges, most readily arise: something just doesn't smell right. We can't quite put our finger on the problem, because most pheromones have no odour and we are unconscious of receiving them at all. And it is a very short step from feeling uncomfortable for no reason to blaming someone else for our unease.

On the positive side, the same odours, permeating everything, are a large part of what makes a house a home. They are an integral part of the bonding process, and we miss them when they are not around. So we carry something familiar away with us. A sweater which once belonged to Dad, a handkerchief of Mum's, a fragment of security blanket, savoured not so much for its texture as for its familial fragrance. And we protest vigorously when anyone tries to launder such a keepsake, not because we can't be parted from it, but because we know from experience that when it comes back, it just won't smell the same.

Our noses never stop. Awake or asleep, we are besieged by smells and have to pick and choose between them, concentrating on the ones that please us, or seem to demand our immediate attention.

Corrosive odours, the really dangerous ones like ammonia, accost us physically, attacking the trigeminal nerve and producing a reflex head movement that jerks our nose back out of the harmful stream. But those that are merely acrid, warning of smoke or fire, are so heavy with hydrocarbons that they impinge also on Jacobson's Organ, setting off an alternative alarm just in case the first one failed to come to our attention.

These are our front-line troops, the first lines of defence that initiate responses appropriate to the occasion. We raise our heads, sniffing and turning from side to side, range-finding, working out distance and direction, making sense out of sensation. Which is all good stuff, but way back behind all this overt activity, I suspect that we have another, more subtle network in play – a fail-safe system that deals with subliminal signals, many of them phero-monal, and is therefore bound, by its very nature, to involve Jacobson's Organ.

This is the realm of intuition, of good and bad 'vibes' and sneaking feelings. The threshold levels here are so low that we are entering the world in which eel larvae smell their way across entire oceans, male moths respond to single molecules, and even bloodhounds may give up in despair. We are talking about stimuli so slight that nothing but a living multi-sensory system will ever be able to detect them, even though it may not know it is doing so.

For instance, if I am right about *Lophiomys* being able to unsettle potential predators by making them feel uneasy, then who knows how many other species practise such psychological deterrence – including our own? We may even do it rather a lot, quite unconsciously, sending out olfactory signals which convey disappointment, disbelief, disdain, distrust or outright disgust. These are all negative, but often necessary ways we have of distancing ourselves from others without making an issue of it, a little like plants using chemicals to inhibit the growth of others nearby.

I have no idea how extensive this no-go area may be. If it is set up by relatively large, less volatile molecules – the ones most likely to impinge on Jacobson's Organ – then it cannot extend for more than a metre or two. That is probably all that is necessary anyway to prevent the participants from coming to blows. It is likely that the finer nuances of such dissociation are carried by combinations of smell and visual or vocal signals, but I suggest that we do have a short-range body odour which is truly deterrent. Not just the normal musky smell from underarm secretions, but something perhaps derived from them to become more skunky, not in odour but in action, something that warns others around us that we are unhappy with a situation. There are several variations on this theme.

In the basic situation, one person, without any very good or immediate reason, decides that they want to be alone. At low levels, this works well. A wall of odourless chemical goes up, and the result is that the seat next to them in the cafeteria or on the train never gets occupied, or not for very long. But if the signal is too strong or becomes prolonged, it can backfire and become part of the 'Garbo syndrome'. The effect of this is universal: everyone takes it personally. No one likes to be dismissed without being given a chance, and sooner or later the source is not just left alone, but completely ostracized for being olfactorily incorrect.

The second situation is bilateral, usually benign but sometimes dangerous. It arises when two people, for whatever reason, find themselves having to associate against their will, perhaps because of a work assignment or simply because they have been introduced by mutual friends. Then both bridle, both signal, and the resulting interference pattern compounds the problem, turning mutual antipathy into outright, irrational dislike. More often than not the situation is resolved by parting, but if that cannot be arranged, things may get nasty. It is fortunate, perhaps, that most of those who suffer from 'road rage' are olfactorily isolated in closed or air-conditioned cars.

And finally, and happily rarely, there is the extreme variation of scent deterrence. *Lophiomys* seems to have gone over the top in this respect, but there may be humans with the same condition. It is still dissociative, but now lacks any hint of social concern. It is manifest by someone in whom disdain has turned to contempt, someone who regards everyone else as irrelevant and disposable – someone psychotic, on whose skin such a signal would be strong enough and distinct enough to be detectable at a distance, perhaps even with the source out of sight. It would be a long-range signal that acts on the olfactory epithelium and warns of someone so strange, so out of control, that they cannot be ignored. The scent of something alien.

This is perhaps what Tonto means when he stops his horse, flares his nostrils, and tells the Lone Ranger that something doesn't 'smell right'. And sure enough, the really bad guys, the ones without hats at all, are invariably lying in wait, in ambush, just around the bend.

It is possible to know things we shouldn't know. 'Intuition' may be no more than an idea which rises, without bidding, to conscious attention. It appears to originate in the unconscious, but its origin is often olfactory. Perhaps it arrives in the limbic system via Jacobson's Organ, as a direct result of 'inspiration'.

Helen Keller articulated her sense of impending weather as 'a throb of expectancy', an elegant way of describing the action of ionized air on the nasal mucosa before a thunderstorm. This is an intuition more climatic than clairvoyant. But in 1780, an obscure civil servant on Île de France (now Mauritius) wrote to his Minister of Marine and announced that he had discovered a way of detecting ships while they were still below the horizon. He called it *nauscopie*, and described it as 'anticipating by means of smell', the olfactory equivalent of 'far sight', a sort of 'far smell'. Unfortunately we know no more details about his technique, but history records that it enabled him to win a great deal of money in wagers.

At a purely mechanistic level, we know that a moth can be aware, by smell alone, of a potential mate many kilometres away. Some migratory birds can probably do as well, navigating ocean legs of their journey at least in part by the smell of land a long way upwind. And gifted humans, like many of those handicapped by the loss of sight or hearing, have shown that they are able to smell, and recognize the smell, of another human well out of sight and beyond hearing.

But I believe that there may be still other ways of olfactory knowing. These involve synesthesia, not taking away a sense and allowing others to compensate for the loss, but adding one sense to another in ways that reinforce them both. If you are a passenger in a car on a long journey, it is very easy to succumb to the soothing and repetitive sounds and images by falling asleep. If you are driving, you run some of the same risk, but fortunately the motor skills required for being at the wheel demand a level of brain and body activity that not only keeps you awake, but allows your mind the luxury of free association. Some of us do our best thinking while driving to and from work. Being driven doesn't work nearly as well.

So, I suggest that any activity which involves the use of vision and hearing, or touch and taste, lifting body functions above a certain critical level, enhances our ability to respond to smell. A hunter walking through the woods, alert to every sound, eyes scanning for the least sign of movement, is going to notice an elusive hint of smoke in the air long before his companion sitting quietly back at camp. This is a well-known phenomenon. It is called sensory facilitation and doesn't lower our threshold for any stimulus, but it does help to bring that information to conscious attention. And if the news is odorous, it also helps to be exposed to it at the right time.

The three-hour cycle of alternation between left and right nostrils goes on night and day. At night, it contributes to sleep movements, which ensure that we roll over onto our left side when the right nostril is dominant, and vice versa. But when we are awake and conscious, the flow of olfactory information travels more to one side of the brain than the other.

There is evidence that our left cerebral lobe is more logical and analytic. It knows the names of smells. The right lobe is more intuitive and emotional. It sniffs things out and has feelings about them. Ideally, we need both kinds of input to come to any useful conclusion. But if a situation is strange and requires action based more on prediction than precedent, you would be better off facing it with a clear left nostril.

Clairvoyance is usually described as mental 'seeing' or a way of knowing what exists 'out of sight'. A clairvoyant is said to be someone who has exceptional 'insight'. The emphasis, as usual, is a visual one, but I suspect that the information garnered is far more often olfactory.

Smell is a long-distance and long-lasting sense. It is more persistent and far less time-dependent than either sight or sound. Odours linger, providing information after the event and offering hints of things yet to come. Our hunter needs to pick up only the smallest subliminal suggestion of lion scent, something that gets into his left nostril and tickles both olfactory epithelium and Jacobson's Organ before stimulating the right brain in a way that whispers 'lion' and allows the hunter to act on a 'hunch'. Then his left brain gets involved and he changes his mind, deciding, for no reason he can justify logically, not to go down to the waterhole after all, but to circle around and take a more careful look from a nearby hill. From there, he sees the lion lying in wait, right beside the path he would have taken.

Clairvoyance? Not really. He had no glimpse of the unknown, no flash of the future. He was in no position to know what was going to happen. But what he did have was a tiny clue, an unconscious hint of what might happen *if* he carried on. All it took was a molecule or two, but that was enough to set in train a line of action that saved his life. It produced a response with unquestionable survival value – and nature encourages things like that and sees to it that they happen again.

We can 'smell a rat' and sense that something is wrong. And we can, equally easily and with the same equipment, feel certain that things are going right. Newborn infants smile the very first time they recognize their mother's smell. This reaction seems to be innate, and certainly encourages most mothers to go on making their odour available. So what we have going between mother and child is a system of mutual support and encouragement, reinforcing the perception of some smells in the baby's life as 'pleasant'.

We learn as we go along. It seems, however, that other smells may be intrinsically pleasurable, not because we know and like particular fragrances which we associate with happy times, but because they are pheromonal and strike an ancient chord in our minds. More often than not these trigger scents are unfamiliar and seem to appeal to us because they are just strange enough to be novel, but not so strange that they make us anxious. So it is possible to fall in love simply because an odour resonates with something in you. It may stimulate one of the limbic pleasure centres which produce natural amphetamines, such as phenylethylamine, that act on the brain like cocaine: nature's way of bringing joy and sex into our lives.

When the state of euphoria we call 'falling in love' abates, as it must after a long period of unconstrained passion, there are other designer scents on hand to reinforce a wobbly pair bond. One of the best known is the scent that comes from the crown of a baby's head of hair. This is secreted and lingers there from the very first day of birth, and is one of the natural opiates or endorphins that promotes a relaxed, easygoing feeling of well-being, calming all anxiety. It does this by triggering production of a hormone called oxytocin, which promotes maternal and paternal behaviour and makes the couple a lot more comfortable with each other after the first fine madness begins to fade. And every part of such subliminal control is mediated by and through Jacobson's Organ.

There are still those who deny the existence of human pheromones. But many of the same people go to extraordinary lengths to suppress their production and distribution. Many religions forbid dancing altogether, or limit it to same-sex couples. Some require women to cover their hair with a *chador* or *hajib*, effectively isolating odours produced by the scalp. Not to do so, they say, dishonours the husband, but it is clearly not *his* honour that is seen to be at risk. The lifting of the bridal veil in Christian marriage ceremonies repeats the same concerns, celebrating the union, giving permission for the oral exchange of pheromones in a ritual kiss.

But everywhere else, olfactory intimacy is frowned upon and severely curtailed by shaving off underarm hair and the thorough deodorizing of all fragrant sources. Why else would puritans insist on high collars, buttoned-up formality, long sleeves, and tight wrist-cuffs? And why, even in our most liberated societies, do we still fuss about keeping nipples covered? It wouldn't have anything to do with the fact that female aureolas are rich in apocrine glands and sex attractants, would it?

And is it just coincidental that male and female circumcision targets two of the most productive pheromonal areas on the body? There is nothing in the Bible or the Koran which calls for such surgery. And there is a nice irony in the fact that circumcision ceremonies, almost everywhere they occur, are accompanied by the pheromonal odours of carefully selected plants. Once again, our actions reveal the existence and potency of the very things we seek to deny.

Most other mammals are clearly aware of the smell of urine and the pheromonal information it contains. We have been socialized and conditioned to ignore the scent of urine, even behaving as though it were repulsive, especially in urban settings. But experiments with human urine have shown that it contains the same active ingredients as most other warm-blooded animals, and has some of the same effects on them as their own urine.

The scent of urine from a female mouse lessens the risk of attack from strange males. Females can wander through the territory of most male mice with impunity, and a male masked with female odour enjoys the same immunity – even if the urine comes from a female human. On the other hand, urine from a male mouse sends a very different message. It can be attractive to female mice, but it predisposes other males to be aggressive, or at least to perceive its presence as an aggressive signal. Urine from a human male produces similarly aggressive behaviour in male mice.

Male rabbits urinate on their females before copulation, and this scent mark tends to repel other males. In the light of this we should look again at the custom amongst human males of giving their wives and lovers gifts of perfumes containing androgenic pheromones. And it raises fascinating speculations about the role of urinary odours on aggression between men. The most hostile graffiti are always to be found in men's locker rooms and lavatories. English football clubs know the risks, and take unusual care to ensure that male supporters from rival clubs are never confronted with each other's territorial toilet odours. They get separate facilities.

We know that the odour of a strange male's urine lowers the level of pituitary hormones in pregnant female mice and results in miscarriages. It has been shown that strange male urine can even shift the hormonal balance so far that female hamsters never become pregnant at all. And the same results in rodents have been produced by androgen-rich human urine. But no one seems to be looking at the possible role of smell on the fertility of women who work, perhaps as cleaners, in surroundings where they may be exposed to the urinary odours of unfamiliar men.

We should worry about such things. They represent real risks to any woman who is both pregnant and has an intact Jacobson's Organ. An informal survey among my own friends suggests that young wives do often have difficulty conceiving for the first time when living in the home of their husband's family, surrounded by the odours of brothers and fathers-in-law. They often fall pregnant within months of finding a home of their own.

Human behaviour is usually more complex and difficult to analyse than that of other species. It is never quite as straightforward in its connection between stimulus and response, or as hard-wired as that of species whose behaviour is largely instinctive. But many of our reactions, particularly those influenced by subliminal signals, remain automatic and unthinking. And these continue to have a significant effect on everyday interactions.

We know that pheromones can normalize irregular menstrual cycles, suppress ovulation and synchronize biorhythms in both mixed- and same-sex couples. And most of these effects seem to come about as a result of interaction with hormone production in the pituitary gland and the hypothalamus. That takes time, but there are at least as many direct connections between outside odours and those areas of the brain that produce rapid shifts in attitude and emotional response. This is the sphere of instant dislikes and sudden, inexplicable aversions.

Do you suddenly find yourself at odds with a waiter or someone serving you at a shop counter, without knowing why? Has anyone you passed in the street ever left a shiver going down your spine, as though you have just had a brush with evil? Can you tell, even with your back turned, that the person behind you in the queue is unstable and likely to do something embarrassing or even dangerous? Are you sometimes able to tell when someone you have never met before is lying to you? Or has someone to whom you have just been introduced made it quite clear that they don't like you? Of course these things have happened to you. They are all part of our normal, if not everyday experience. We are finely tuned receivers, adept at detecting nuances of anything unusual in others, because our lives and livelihood depend on it.

Natural selection works that way. We look and listen very hard, picking up information unconsciously all the time and justifying our reaction by saying, 'I don't like the look of that' or 'I didn't like the sound of it at all.' Often we invoke our sense of touch and say, 'I can't quite put my finger on it, but . . .' True to our usual sensory bias, we give credit to our sense of smell only as a last resort, when something is so obviously wrong that we can say, with conviction, that 'it stinks!' The chances are, however, that what actually alerted us right from the start was a smell. More often than not it is a smell without perceptible odour, one capable of crossing large spaces in vanishingly small amounts of time, and changing our minds with the impact of nothing more than a scatter of molecules on Jacobson's Organ.

Even among scientists there is a tendency to denigrate our sense of smell by classifying all primates as microsmatic – as 'poor smellers'. It is true that a relatively small area of our brain is given over to the olfactory sense, and that we have fewer small receptor cells than most other mammals. Rabbits have fifty million receptors, monkeys seldom more than ten million. But this ignores the fact that receptor cells are not necessarily finely tuned to particular odours. They can respond to a wide range of odours, making it possible for a few cells to do a lot of different things.

It also ignores the neurological fact that much olfactory information is encoded in ways that require recognition in the brain. So the bigger the brain, the more sense it can make of limited information. This gives monkeys, apes and humans the chance to demonstrate a remarkably sophisticated sense of smell. And that opens the door to an unexpected conclusion: far from being poor smellers, we may in some ways be the most evolved of all species in this respect.[96]

In the right circumstances, it may even be possible for us to use our sense of smell, with the help of Jacobson's Organ, to find out:

Whether or not it will rain.

If there really is a snake under the porch.

When the figs are ripe on that tree down at the river.

Who is coming up through the orchard.

Where we left the car keys.

Which way the children went.

Who their friends are.

Who last used this chair or slept in that bed, and whether they were alone.

When the girl next door ovulates and is likely to be attractive to, or a threat to others.

What our spouses had for lunch, and who they spent that time with.

And whether or not we are going to need a lawyer.

There is nothing paranormal about any of these intuitions, but they would be well within the capability of a sense of smell expanded in ways that make it more like a 'sixth sense'.

Most discussions of pheromones in humans are limited to the subject of inducing sex-related behaviour in potential mates. The fact that it is the Erox Corporation which is paying for and promoting some of the most interesting work means that the emphasis for the moment is on sensual attractants in fragrant form, but I believe that Jacobson's Organ is involved in far more than our sex lives.

Humans have no distinct pattern of behaviour for distributing hormones outside the body. We don't as a rule urinate on lamp-posts, or sniff each other's ankles, or spray out our scent in a fragrant cloud. But we do carry complex odours on our skins and leave a vapour trail of scented cells behind us wherever we go. These slough off at the rate of ten thousand cells an hour from every finger, creating an invisible but highly individual cloud of volatile and odourless particles. Some of these are short-lived; others last longer. A few may be very persistent and, if they are not consumed by mites and are protected from self-destruction by a hot, dry climate, they could survive for centuries. Who knows what our noses may make of them then?

The message may be intact, even in a single discarded cell, but it was always low on characterization, something like a classified ad:

> Young white male, athletic, non-smoker,
> reasonably cheerful, fond of garlic.

And that much could still be available to a bloodhound, or to any human with a good nose and an active social life. There is nothing like the richness and detail of the madeleine effect: this is not memory, but there could be sufficient information anyway to trigger some synesthetic effect, enough to dress up the illusion, perhaps even in period costume, as a fully fledged 'ghost'.

I have tried for years to make sense of ghosts. Most are sedentary – they seem to be fixed in a particular location. All are clothed, which is hard to explain, even if you can accept the possibility of spirit survival. And many do indeed come complete with a characteristic smell. That encourages me, and may be a useful clue. Smell may be all that connects us to the provider of the stimulus, the previous owner of the pheromonal cells. Smell is a potent channel of communication, but its vocabulary is rather limited and provides us with only just enough detail for a hazy outline, a very sketchy biography, and limited intelligence – which is actually a very good description of most spirits. They are high on emotional impact, low on useful facts about the great beyond. Everything you would expect from a simple molecular message or an olfactory telegram. It was Plato who called odour a 'half-formed nature'.

But if the cells were once attached to someone you knew personally, the result could be very different. The smell of the cell itself could be your madeleine. You could find yourself confronted, not just by a vague impression, but by a full synesthetic symphony of all the senses, including words and pictures, tastes and textures, everything necessary for a complex and completely convincing illusion: a classic haunting. At a lower level, this explains the persistent feeling among those recently bereaved that their loved ones are still around them.

I don't believe we have even begun to come to terms with our 'sense of wonder'. We still have a lot to learn, particularly from those parts of the sense housed in the rhinencephalon – the old 'smell brain'. Mysteries abound there, and reveal themselves only when we have the courage and the imagination to ask the proper questions. Jean Jacques Rousseau said, 'The sense of smell is the sense of imagination. It greatly disturbs the brain.'[156] Indeed, and it still does.

And I can do little better than to leave the last word to a poet-philosopher who had more courage than most. In his impassioned *Twilight of the Idols*, Friedrich Nietzsche said:

In the quest for truth, the sense of smell – which is also the sense of veracity, drawing as it does upon the sure sources of animal instinct that give the body its great wisdom, providing the tool for a psychologist in search of the fake and the illusory – dethrones the chilly logic that emerges when man struggles against the intellectual. Above and beyond its primary function, smell thus serves as a 'sixth sense', the sense of intuitive knowledge.[135]

So be it.

I should think we might fairly gauge the future of biological science, centuries ahead, by estimating the time it will take to reach a complete, comprehensive understanding of odour. It may not seem a profound enough problem to dominate all the life sciences, but it contains, piece by piece, all the mysteries.

LEWIS THOMAS, 'On Smell' in *Late Night Thoughts*

Taxonomy

The species of plants and animals mentioned in the text by their common names are here identified more precisely by their Latin binomial names, which Carolus Linnaeus would have recognized. Domesticated breeds and species mentioned in passing are not included.

FUNGUS
Truffle   *Tuber melanosporum*

PLANTS

Acacia	*Acacia* sp.
Agrimony	*Agrimonia parviflora*
Catnip	*Nepeta cataria*
Common poplar	*Populus euroamericana*
European walnut	*Juglans regia*
Garlic	*Allium sativum*
Goldenrod	*Solidago ulmiflora*
Laurel	*Laurus nobilis*
Madagascar periwinkle	*Catharanthus roseus*
Myrtles	*Myrtus communis*
Norway spruce	*Picea abies*
Onycha	*Cistus ladniferus*
Pine	*Pinus* sp.
Red alder	*Alnus rubra*
Rock cress	*Arabidopsis thaliana*
Rose	*Rosa* sp.

Rosemary	*Rosmarinus officinalis*
Sensitive mimosa	*Mimosa pudica*
Sitka willow	*Salix sitchensis*
St John's wort	*Hypericum* sp.
Subterranean clover	*Trifolium subterraneum*
Sugar maple	*Acer saccharum*
Tobacco	*Nicotiana tabacum*
Yarrow	*Achillea millefolium*

INSECTS

Carpenter ant	*Camponotus socius*
Silkworm moth	*Bombyx mori*
Tent caterpillar	*Malacosoma californica*

SPIDER

Bolas spider	*Mastophora* sp.

CRUSTACEAN

Brime shrimp	*Artemia salina*

FISH

European eel	*Anguilla anguilla*
European minnow	*Phoxinus phoxinus*
Spotted dogfish	*Scyliorhinus caniculus*
White-tip shark	*Triaenodon obesus*

AMPHIBIANS

African clawed toad	*Xenopus laevis*
American leopard frog	*Rana pipiens*
Common toad	*Bufo bufo*
European common frog	*Rana temporaria*
Mexican toad	*Bufo valliceps*
Red-bellied newt	*Taricha rivularis*
Spotted chorus frog	*Pseudacris clarkii*
Strecker's chorus frog	*Pseudacris streckeri*

REPTILES

Plains garter snake	*Thamnophis radix*
Western whiptail lizard	*Cnemidophorus tigris*

BIRDS

Antarctic snow petrel	*Pagodroma nivea*
European starling	*Sturnus vulgaris*
Greater shearwater	*Puffinus gravis*
Turkey vulture	*Cathartes aura*
Wilson's petrel	*Oceanites oceanicus*

MAMMALS

African lion	*Panthera leo*
Black rhinoceros	*Diceros bicornis*
Boar	*Sus scrofa*
Bonobo	*Pan paniscus*
Brown hyena	*Hyaena brunnea*
Chimpanzee	*Pan troglodytes*
Civet	*Viverra sp.*
Common vampire bat	*Desmodus rotundus*
Crested 'rat'	*Lophiomys imhausi*
European badger	*Meles meles*
European hedgehog	*Erinaceus europaeus*
European rabbit	*Oryctolagus cuniculus*
European weasel	*Mustela nivalis*
Golden hamster	*Mesocricetus auratus*
Green acouchi	*Myoprocta prattii*
Hippopotamus	*Hippopotamus amphibius*
House mouse	*Mus musculus*
Indian elephant	*Elephans maximus*
Jaguar	*Panthera onca*
Mongolian gerbil	*Meriones unguiculatus*
Mule deer	*Odocoileus hemionus*
Musk deer	*Moschus moschiferus*
Prairie vole	*Microtus ochrogaster*
Pronghorn	*Antilocapra americana*

Ring-tailed lemur	*Lemur catta*
Roe deer	*Capreolus capreolus*
Short-tailed vole	*Microtus agrestis*
Spiny mouse	*Acomys caharinus*
Spotted hyena	*Crocuta crocuta*
Springbok	*Antidorcas marsupialis*
Striped skunk	*Mephitis mephitis*
Sugar glider	*Petaurus breviceps*
Water vole	*Arvicola terrestris*
Weddell seal	*Leptonychotes weddelli*
White-bellied shrew	*Suncus murinus*
White-tailed deer	*Odocoileus virginianus*

A Note on the Illustrations

All the botanical illustrations in the text are taken from woodcuts made by Carolus Clusius (Charles de l'Ecluse) for his pioneering publication of *Rariorum plantarum historia* in Antwerp in 1601.

Clusius was as influential in establishing modern botany as Linnaeus, a century later, was in providing the foundation for modern taxonomy.

Bibliography

1. Ackerman, Diane. *A Natural History of the Senses*. New York: Chapman (1990).
2. Adams, Donald R., and Michael D. Wiekamp. 'The canine vomeronasal organ'. *Journal of Anatomy*, *138*, 771 (1984).
3. Adams, M. D. 'Seasonal changes in the skin glands of roe deer'. Ph.D. Thesis, Reading University (1976).
4. Adams, M. G. 'Odour producing organs of mammals'. *Symposia of Zoological Society of London*, *45*, 57 (1980).
5. Agosta, William C. *Chemical Communication: The Language of Pheromones*. New York: Scientific American Library (1992).
6. Audubon, John J. 'Account of the habits of the turkey vulture'. *Edinburgh New Philosophy Journal*, 2, 172 (1827).
7. Ayabe-Kanamura, Saho, Ina Schicker, Mattias Lasker, Robyn Hudson, Hans Distel, Tatsu Kobayakawa and Sachiko Saito. 'Differences in perception of everyday odors: A Japanese–German cross-cultural study'. *Chemical Senses*, *23*, 31 (1998).
8. Baldwin, B. A., and Elizabeth E. Shillito. 'The effects of ablation of the olfactory bulbs on parturition and maternal behaviour in Soay sheep'. *Animal Behaviour*, 22, 220 (1974).
9. Baldwin, Ian T., and Jack C. Schultz. 'Rapid changes in tree leaf chemistry induced by damage'. *Science*, *221*, 277 (1983).
10. Bang, Betsy G., and Stanley Cobb. 'The size of the olfactory bulb in 108 species of bird'. *The Auk*, *85*, 55 (1968).
11. Bang, Betsy G. 'Functional anatomy of the olfactory system in 23 orders of birds'. *Acta Anatomica*, *79*, Sup. 58, 1 (1971).

12. Berliner, David L., Clive Jennings-White and Robert M. Lavker. 'The human skin: Fragrances and pheromones'. *Journal of Steroid Biochemistry and Molecular Biology*, 39, 671 (1991).

13. Berliner, David L., Luis Monti-Bloch, Clive Jennings-White and Vicente Diaz-Sanchez. 'The functionality of the human vomeronasal organ (VNO): Evidence for steroid receptors'. *Journal of Steroid Biochemistry and Molecular Biology*, 58, 26, (1996).

14. Brill, A. A. 'The sense of smell in the neuroses and psychoses'. *Psychoanalysis Quarterly*, 1, 7 (1932).

15. Brockie, R. 'Self-anointing by wild hedgehogs'. *Animal Behaviour*, 24, 68 (1976).

16. Broom, Robert. 'A contribution to the comparative anatomy of the mammalian Organ of Jacobson'. *Transactions of the Royal Society of Edinburgh*, 39, 231 (1897).

17. Broom, Robert. *The Mammal-Like Reptiles of South Africa*. London: Witherby (1932).

18. Broom, Robert. 'On the palate, occiput and hindfoot of *Bauria cynops*'. *American Museum Novitates*, No. 946, 106 (1937).

19. Brownlee, Robert G., and Robert M. Silverstein. 'Isolation, identification and function of the chief component of the male tarsal scent in black-tailed deer'. *Nature*, 221, 284 (1969).

20. Burger, J., and M. Gochfeld. 'A hypothesis on the role of pheromones on age of menarche'. *Medical Hypotheses*, 17, 39 (1985).

21. Butenandt, A., R. Beckmann, D. Stam and E. Hecker. 'Über den Sexual-Lockstoff des Seitenspinners *Bombyx mori*'. *Zeitschrift für Naturforschung*, 14, 283 (1959).

22. Cain, William S. 'Educating your nose'. *Psychology Today*, 15, 48 (1981).

23. Cain, William S. 'What do we remember about odours?' *Perfumer and Flavorist*, 9, 17 (1984).

24. Carson, Rachel. *The Sea Around Us*. London: Staples Press (1951).

25. Cernoch, Jennifer M., and Richard H. Porter. 'Recognition of maternal axillary odors by infants'. *Child Development*, 56, 1593 (1985).

26. Charles, Dan. 'Stressed plants cry for help'. *New Scientist*, 8 March, p. 7 (1997).

27. Clark, Larry, and J. Russell Mason. 'Olfactory discrimination of plant volatiles by the European starling'. *Animal Behaviour*, 35, 227 (1987).

28. Classen, Constance, David Howes and Anthony Synnott. *Aroma: The Cultural History of Smell*. London: Routledge (1994).

29. Classen, Constance. *Worlds of Sense: Exploring the Senses in History and Across Cultures*. London: Routledge (1995).

30. Comfort, Alex. 'Likelihood of human pheromones'. *Nature*, 230, 432 (1971).

31. Conniff, Richard. 'We shed 50 million skin cells a day'. *Smithsonian*, January, p. 65 (1986).

32. Cooper, J. Gregory. 'Comparative anatomy of the vomeronasal organ complex in bats'. *Journal of Anatomy*, 122, 571 (1976).

33. Corbin, Alain. *The Foul and the Fragrant: Odor and the French Social Imagination*. Cambridge: Harvard University Press (1986).

34. Cowley, J. J., and B. W. L. Brooksbank. 'The effect of two odorous compounds on performance in an assessment-of-people test'. *Psychoneuroendocrinology*, 2, 159 (1977).

35. Cowley, J. J., and B. W. L. Brooksbank. 'Human exposure to putative pheromones and changes in aspects of social behaviour'. *Journal of Steroid Biochemistry and Molecular Biology*, 39, 647 (1991).

36. Crews, David. 'Hormonal control of courtship behavior in the garter snake'. *Hormones and Behavior*, 7, 451 (1976).

37. Crews, David, and William R. Garstka. 'The ecological physiology of a garter snake'. *Scientific American*, 247, 159 (1982).

38. cummings, e.e. *Selected Poems*. London: Faber (1969).

39. Dabney, V. 'Connections of the sexual apparatus with the ear, nose and throat'. *New York Medical Journal*, 97, 533 (1913).

40. Day, Stephen. 'The sweet smell of death'. *New Scientist*, 7 September, p. 28 (1996).

41. Dole, J. W. 'The role of olfaction in the orientation of leopard frogs'. *Herpetologica*, 28, 258 (1972).

42. Doty, Richard L. 'Changes in the intensity and pleasantness of human vaginal odors during the menstrual cycle'. *Science*, 190, 1316 (1975).

43. Doty, Richard L. 'Olfactory communication in humans'. *Chemical Senses*, 6, 351 (1981).

44. Doty, Richard L. 'Communication of gender from human breath odors'. *Hormones and Behavior, 16*, 13 (1982).

45. Durden-Smith, Jo, and Diane deSimone. *Sex and the Brain*. New York: Arbor House (1983).

46. Dryden, G. L., and C. H. Conway. 'The origin and hormonal control of scent production in *Suncus murinus*'. *Journal of Mammology, 48*, 420 (1967).

47. Duvall, David. 'A new question of pheromones'. In Ref. 78.

48. Empson, J. 'Periodicity in body temperature in man'. *Experientia, 33*, 342 (1977).

49. Estes, R. D. 'The role of the vomeronasal organ in mammalian reproduction'. *Mammalia, 36*, 315 (1972).

50. Ewer, R. F. *Ethology of Mammals*. London: Elk Press (1968).

51. Fabricant, N. D. 'Sexual functions and the nose'. *American Journal of Medical Science, 239*, 156 (1960).

52. Farbman, Albert I. *Cell Biology of Olfaction*. Cambridge: Cambridge University Press (1991).

53. Finger, Thomas E., and Wayne L. Silver. *Neurobiology of Taste and Smell*. New York: Wiley (1988).

54. Finlay, G. H. *Dr. Robert Room: Palaeontologist and Physician, 1866–1951*. Cape Town: Balkema (1972).

55. Fisher, H. E. *The Anatomy of Love: A Natural History of Mating, Marriage, and Why We Stray*. New York: Norton (1992).

56. Foster, Steven, and James A. Duke. *A Field Guide to Medicinal Plants: Eastern and Central North America*. Boston: Houghton Mifflin (1990).

57. Fowles, John. 'Seeing nature whole'. *Harpers*, November, p. 49 (1979).

58. Freitag, Joachim, Jürgen Krieger, Jörg Strotmann and Heinz Breer. 'Two classes of olfactory receptors in *Xenopus laevis*'. *Neuron, 15*, 1383 (1995).

59. Freud, Sigmund. 'Notes upon a case of obsessional neurosis'. In *The Standard Edition of the Complete Psychological Works*, Vol. 10. London: Hogarth Press (1953).

60. Garcia-Velasco, Jose, and Manuel Mondragon. 'The incidence of the vomeronasal organ in 1000 human subjects and its possible clinical significance'. *Journal of Steroid Biochemistry and Molecular Biology, 39*, 561 (1991).

61. Gemme, R., and C. C. Wheeler (eds.). *Progress in Sexology: Selected Papers from the Proceedings of the 1976 International Congress of Sexology*. New York: Plenum (1977).

62. Gibbons, Boyd. 'The intimate sense of smell'. *National Geographic*, September, p. 324 (1986).

63. Godfrey, J. 'The origin of sexual isolation between bank voles'. *Proceedings of the Royal Physical Society of Edinburgh*, 47, (1958).

64. Good, Paul R., Nori Geary and Trygg Engen. 'The effect of oestrogen on odour detection'. *Chemical Senses and Flavour*, 2, 45 (1976).

65. Gorman, Martyn L. 'Sweaty mongooses and other smelly carnivores'. In Ref. 185.

66. Graham, C. A., and W. C. McGrew. 'Menstrual synchrony in female undergraduates living on a coeducational campus'. *Psychoneuroendocrinology*, 5, 245 (1980).

67. Grant, E., O. Anderson and V. C. Twitty. 'Homing orientation by olfaction in newts'. *Science*, 160, 1354 (1968).

68. Grubb, T. C. 'Smell and foraging in shearwaters and petrels'. *Nature*, 237, 404 (1972).

69. Grubb, Jerry C. 'Olfactory orientation in *Bufo woodhousei fowleri*, *Pseudacris clarki* and *Pseudacris streckeri*'. *Animal Behaviour*, 21, 726 (1973).

70. Grubb, Jerry C. 'Maze orientation in Mexican toads'. *Journal of Herpetology*, 10, 97 (1976).

71. Halpern, Mimi, and John L. Kubie. 'Chemical access to the vomeronasal organs of garter snakes'. *Physiology and Behavior*, 24, 367 (1980).

72. Halpern, Mimi, and John L. Kubie. 'The role of the ophidian vomeronasal systems in species-typical behavior'. *Trends in Neurological Science*, 7, 472 (1984).

73. Hart, Benjamin L. 'Flehmen behaviour and vomeronasal organ function'. In Ref. 133.

74. Hasler, A. D., and James A. Larson. 'The homing salmon'. *Scientific American*, 202, 72 (1958).

75. Hedin, P. A. (ed.). *Plant Resistance to Insects*. American Chemical Society Symposia, Vol. 208 (1982).

76. Holldobler, Burt, and Edward O. Wilson. *Journey to the Ants: A*

Story of Scientific Exploration. Cambridge: Harvard University Press (1994).

77. Hooley, Richard. *Current Biology*, in the press (1999).

78. Hotton, Nicholas III (ed.). *Ecology and Biology of Mammal-like Reptiles*. Washington: Smithsonian Institution (1986).

79. Houston, David C. 'Scavenging efficiency of turkey vultures in tropical forest'. *Condor*, 88, 318 (1986).

80. Hudson, R., and H. Distel. 'The pattern of behaviour of rabbit pups in the nest'. *Behaviour*, 79, 255 (1982).

81. Hutchinson, L. V., and Bernice M. Wenzel. 'Olfactory guidance in foraging by Procellariiformes'. *Condor*, 82, 314 (1980).

82. Huysmans, J. K. *Against the Grain*. New York: Dover (1969).

83. Jackson, D. D. 'Psychotherapy for schizophrenia'. *Scientific American*, 188, 58 (1953).

84. Jacobson, L. 'Description anatomique d'un organe observé dans les mammifères'. *Annales Musée Histoire Naturelle*, 18, 412 (1811).

85. Jellinek, J. S., B. Olies du Bosque, J. Gschwind, B. Schubert and A. Scharf. 'The scent and the marketing mix'. *Dragoco Report*, 39, 103 (1992).

86. Jellinek, J. S. 'Perfumes as signals'. *Advances in the Biosciences*, 93, 585 (1994).

87. Jellinek, Paul. *The Practice of Modern Perfumery*. New York: Interscience (1954).

88. Jennings-White, Clive. 'Perfumery and the sixth sense'. http://www.erox.com/SixthSense/StoryOne.html

89. Johnston, R. E. 'Responses to individual signatures in scent countermarks'. *Advances in the Biosciences*, 93, 361 (1994).

90. Jolly, Alison. *Lemur Behavior: A Madagascar Field Study*. Chicago: Chicago University Press (1966).

91. Kalmijn, A. J. 'The electric sense of sharks and rays'. *Journal of Experimental Biology*, 55, 371 (1971).

92. Kalmus, H. 'The discrimination by the nose of the dog of individual human odours and in particular the odour of twins'. *British Journal of Animal Behaviour*, 3, 25 (1955).

93. Karlson, P., and M. Lüscher. 'Pheromones: A new term for a class of biologically active substances'. *Nature*, 183, 55 (1959).

94. Kaufman, G. W., D. B. Siniff and R. Reichle. 'Colonial behaviour

of Weddell seals at Hutton Cliffs, Antarctica'. *Rapports du Conseil Permanent International pour l'Exploration de la Mer*, *169*, 228 (1975).

95. Keller, Helen. 'Sense and sensibility'. *The Century Magazine*, February, p. 66 (1908).

96. Keverne, Eric B. 'Olfaction in the behaviour of non-human primates'. In Ref. 185.

97. Kiltie, Richard A. 'On the significance of menstrual synchrony in closely associated women'. *American Naturalist*, *119*, 414 (1982).

98. Kirk-Smith, Michael, D. A. Booth, D. Carrol and P. Davies. 'Human social attitudes affected by androstenol'. *Research Communications in Psychology, Psychiatry and Behavior*, *3*, 379 (1978).

99. Kirk-Smith, Michael, and D. A. Booth. 'Effect of androstenone on choice of location in other's presence'. In Ref. 196.

100. Kirk-Smith, Michael, S. van Toller and G. H. Dodd. 'Unconscious odour conditioning in human subjects'. *Biological Psychology*, *17*, 221 (1983).

101. Kleerekoper, H., and J. Morgensen. 'Role of olfaction in the orientation of *Petromyzon marinus*'. *Physiological Zoology*, *36*, 347 (1963).

102. Klopfer, P. H., D. K. Adams and M. R. Klopfer. 'Maternal imprinting in goats'. *Proceedings of the National Academy of Sciences*, *52*, 911 (1962).

103. Kobal, G., S van Toller and T. Hummel. 'Is there directional smelling?' *Experientia*, *45*, 130 (1989).

104. Kruczek, M. 'Vomeronasal organ removal eliminates odor preferences in bank voles'. *Advances in the Biosciences*, *93*, 421 (1994).

105. Kruuk, H. 'Spatial organization and territorial behaviour of the European badger'. *Journal of Zoology*, *184*, 1 (1978).

106. Kubie, John L., and Mimi Halpern. 'Laboratory observations of trailing behaviour in garter snakes'. *Journal of Comparative and Physiological Psychology*, *89*, 667 (1975).

107. Kubie, John L., Alice Vagvolgyi and Mimi Halpern. 'Roles of the vomeronasal and olfactory systems in courtship behaviour of male garter snakes'. *Journal of Comparative and Physiological Psychology*, *92*, 627 (1977).

108. Kubie, John L., and Mimi Halpern. 'Garter snake trailing

behaviour'. *Journal of Comparative and Physiological Psychology*, *93*, 362 (1978).

109. Ladewig, Jan, Edward O. Price and Benjamin L. Hart. 'Flehmen in male goats: Role in sexual behavior'. *Behavioural and Neural Biology*, *30*, 312 (1980).

110. Lane, Harlan (ed.). *The Wild Boy of Aveyron*. Cambridge: Harvard University Press (1976).

111. Largey, Gale, P., and David Rodney Watson. 'The sociology of odors'. *American Journal of Sociology*, 77, 1021 (1972).

112. Lawless, Julia. *Aromatherapy and the Mind*. London: HarperCollins (1994).

113. Le Gúerer, Annick. *Scent: The Mysterious and Essential Powers of Smell*. New York: Turtle Bay (1992).

114. Linnaeus, Carolus. *Systema naturae*. Stockholm, 1735.

115. Linnaeus, Carolus. 'Odores medicamentorum'. *Amoenitates Academicae*, *3*, 183 (1756).

116. Maclean, Charles. *The Wolf Children*. New York: Hill & Wang (1978).

117. Maclean, Paul D. 'Neurological significance of the mammal-like reptiles'. In Ref. 78.

118. McClintock, Martha. 'Menstrual synchrony and suppression'. *Nature*, 229, 224 (1971).

119. Marks, Lawrence E. *The Unity of the Senses: Interrelations Among the Modalities*. New York: Academic Press (1978).

120. Martin, L. C. (ed.). *The Poems of Robert Herrick*. Oxford: Oxford University Press (1974).

121. Maugh, Thomas H. 'The scent makes sense'. *Science*, *215*, 1224 (1982).

122. Maugham, W. Somerset. *On a Chinese Screen*. Oxford: Oxford University Press (1985).

123. Mensing, J., and C. Beck. 'The psychology of fragrance selection'. In Ref. 198.

124. Meredith, Michael, and Gordon M. Burghardt. 'Electrophysiological studies of the tongue and accessory olfactory bulb in garter snakes'. *Physiology and Behavior*, 21, 1001 (1978).

125. Meredith, Michael, David M. Marques, Robert J. O'Connell and F. Lee Stern. 'Vomeronasal pump'. *Science*, 207, 1224 (1980).

126. Mertl, A. S. 'Discrimination of individuals by scent in a primate'. *Behavioural Biology*, *14*, 505 (1975).

127. Michael, R. P., and R. W. Bonsall. 'Chemical signals and primate behaviour'. In Ref. 133.

128. Monti-Bloch, Luis, C. Jennings-White, D. S. Dolbert and D. L. Berliner. 'The human vomeronasal system'. *Psychoneuroendocrinology*, *19*, 673 (1994).

129. Moran, David T., Bruce W. Jafek and J. Carter Rowley III. 'The vomeronasal organ in man: Ultrastructure and frequency of occurrence'. *Journal of Steroid Biochemistry and Molecular Biology*, *39*, 545 (1991).

130. Morris, Desmond. *Dogwatching*. New York: Three Rivers Press (1986).

131. Moy, R. F. 'Histology of the forefoot and hindfoot interdigital and medial glands of the pronghorn'. *Journal of Mammology*, *52*, 441 (1971).

132. Muller-Schwarze, D., and R. Silverstein (eds.). *Chemical Signals in Vertebrates*, Vol. 1. New York: Plenum (1977).

133. Muller-Schwarze, D., and R. Silverstein (eds.). *Chemical Signals in Vertebrates*, Vol. 3. New York: Plenum (1983).

134. Müller-Velten, H. 'Über den Angstgeruch bei dem Hausmaus'. *Zeitschrift für vergleichende Physiologie*, *52*, 401 (1966).

135. Nietzsche, Friedrich W. *Twilight of the Idols and How to Philosophize with the Hammer, the Anti-Christ*. Edinburgh: Foulis (1911).

136. Nunley, M. Christopher. 'Response of deer to human blood odor'. *American Anthropologist*, *83*, 630 (1981).

137. Orwell, George. *The Road to Wigan Pier*. London: Gollancz (1937).

138. Owens, M. J., and Delia D. Owens. 'Feeding ecology and its influence on social organization in brown hyenas (*Hyena brunnea*) of the Central Kalahari Desert'. *East African Wildlife Journal*, *16*, 113 (1978).

139. Parsons, Thomas S. 'Nasal anatomy and the phylogeny of reptiles'. *Evolution*, *13*, 175 (1959).

140. Patterson, R. L. S. 'Identification of the musk odour component of boar saliva'. *Journal of Science, Food and Agriculture*, *19*, 434 (1968).

141. Pederson, Patricia E., William B. Stewart, Charles A. Greer and Gordon M. Shepherd. 'Evidence for olfactory function *in utero*'. *Science*, *221*, 478 (1983).

142. Persky, H., H. I. Lief, C. P. O'Brien, D. Strauss and W. Miller. 'Reproductive hormone levels and sexual behaviors of young couples during the menstrual cycle'. In Ref. 61.

143. Pfeiffer, W. 'Alarm substances'. *Experientia*, 19, 113 (1963).

144. Plutarch. *Lives of the Nobel Grecians and Romans*. Boston: Houghton Mifflin (1928).

145. Poddar-Sarkar, M., R. L. Brahmachary and J. Dutta. 'Scent marking in the tiger'. *Advances in the Biosciences*, 93, 339 (1994).

146. Porter, Richard H., and F. Etscorn. 'Olfactory imprinting resulting from brief exposure in *Acomys canirinus*'. *Nature*, 250, 732 (1974).

147. Porter, Richard H., Jennifer M. Cernoch and F. Joseph McLaughlin. 'Maternal recognition of neonates through olfactory cues'. *Physiology and Behavior*, 30, 151 (1983).

148. Porter, Richard H., Jennifer M. Cernoch and F. Joseph McLaughlin. 'Odor signatures and kin recognition'. *Physiology and Behavior*, 34, 445 (1985).

149. Potapov, M. A., and V. I. Evsikov. 'Kin recognition in water voles'. *Advances in the Biosciences*, 93, 247 (1995).

150. Powers, J. Bradley, and Sarah S. Winans. 'Vomeronasal organ: Critical role in mediating sexual behavior of the male hamster'. *Science*, 187, 961 (1975).

151. Preti, G., W. B. Cutler, G. R. Huggins, C. R. Garcia and H. J. Lawley. 'Human axillary secretions influence women's menstrual cycles'. *Hormones and Behavior*, 20, 474 (1986).

152. Proust, Marcel. *Remembrance of Things Past*, Vol. 1. New York: Random House (1981).

153. Riasman, G. 'An experimental study of the projection of the amygdala to the accessory olfactory bulb and its relationship to the concept of a dual olfactory system'. *Experimental Brain Research*, 14, 395 (1972).

154. Rasmussen, Lois E., Michael J. Schmit, Roger Henneous and Douglas Groves. 'Asian bull elephants: Flehmen-like responses'. *Science*, 217, 159 (1982).

155. Rhoades, David F. 'Responses of alder and willow to attack by tent caterpillars and webworms'. In Ref. 75.

156. Rindisbacher, Hans J. *The Smell of Books: A Cultural-Historical*

Study of Olfactory Perception in Literature, Ann Arbor: University of Michigan Press (1992).

157. Roper, Timothy. 'Odour and colour as cues for taste-avoidance learning in domestic chicks'. *Animal Behaviour*, 53, 1241 (1997).

158. Rumbelow, Helen. 'Smell of Blitz brings back memories'. *The Times*, London, 10 October, p. 19 (1998).

159. Russell, Michael J. 'Human olfactory communication'. *Nature*, 260, 520 (1976).

160. Russell, Michael J., Genevieve M. Switz and Kate Thompson. 'Olfactory influences on the human menstrual cycle'. *Pharmacology, Biochemistry and Behaviour*, 13, 737 (1980).

161. Ruysch, Frederick. *Thesaurus anatomicus*. Amsterdam: Wolters (1703).

162. Sanderson, Ivan T. *Living Mammals of the World*. New York: Doubleday (1972).

163. Sartre, Jean-Paul. *Baudelaire*. Paris: Gallimard (1963).

164. Schleidt, M., P. Neumann and H. Morishita. 'A cross-cultural study on the attitude towards personal odours'. *Journal of Chemical Ecology*, 7, 19 (1981).

165. Schmidt, U. and A. M. Greenhall. 'Untersuchungen zur geruchlichen Orientierung der Vampir-fledermaus'. *Zeitschrift für vergleichende Physiologie*, 74, 217 (1971).

166. Schultze-Westrum, T. 'Inneartiche Verstandigung durch Düfte beim Gleitbeutler.' *Zeitschrift für vergleichende Physiologie*, 50, 151 (1965).

167. Seton, E. T. *Lives of Game Animals*. New York: Doubleday (1927).

168. Shelley, W. B., H. J. Hurley, Jr. and A. C. Nichols. 'Axillary odor: Experimental study of the role of bacteria, apocrine sweat and deodorants'. *Archives of Dermatology and Syphiology*, 68, 430 (1953).

169. Shepher, J. *Incest: A Biosocial View*. New York: Academic Press (1983).

170. Shulaev, Vladimir, Paul Silverman and Ilya Raskin. 'Airborne signalling by methyl salicilate in plant pathogen resistance'. *Nature*, 385, 718 (1997).

171. Shutt, Donald A. 'The effects of plant oestrogens on animal reproduction'. *Endeavour*, 35, 110 (1976).

172. Signoret, J. P. 'Chemical communication and reproduction in domestic animals'. In Ref. 132.

173. Simon, Carol A. 'Masters of the tongue flick'. *Natural History*, *91*, 59 (1982).

174. Singer, Alan G. 'A chemistry of mammalian pheromones'. *Journal of Steroid Biochemistry and Molecular Biology*, *39*, 627 (1991).

175. Singh, J. A. L., and R. M. Zingg. *Wolf Children and Feral Man*. New Haven: Archon (1966).

176. Smith, K., and J. O. Sines. 'Demonstration of a peculiar odor in the sweat of schizophrenic patients'. *Archives of General Psychiatry*, *2*, 184 (1960).

177. Smith, R. J. F. 'Alarm substance of fish'. In Ref. 132.

178. Sommerville, Barbara, David Gee and June Averill. 'On the scent of body odour'. *New Scientist*, 10 July, p. 41 (1986).

179. Sonenshine, D. E. 'Pheromones and other semiochemicals of the acari'. *Annual Review of Entomology*, *30*, 1 (1985).

180. Stehn, R. A., and Milo E. Richmond. 'Male induced pregnancy termination in the prairie vole *Microtus ochrogaster*'. *Science*, *187*, 1211, (1975).

181. Stensaas, Larry J. 'Ultrastructure of the human vomeronasal organ'. *Journal of Steroid Biochemistry and Molecular Biology*, *39*, 553 (1991).

182. Stern, Kathleen, and Martha K. McClintock. 'Regulation of ovulation by human pheromones'. *Nature*, *392*, 177 (1998).

183. Stoddart, D. Michael. 'Effect of the odour of weasels on trapped samples of their prey'. *Oecologia*, *22*, 439 (1976).

184. Stoddart, D. Michael. *The Ecology of Vertebrate Olfaction*. London: Chapman & Hall (1980).

185. Stoddart, D. Michael (ed.). *Olfaction in Mammals*. Symposia of the Zoological Society of London, Vol. 45 (1980).

186. Stoddart, D. Michael. 'The role of olfaction in the evolution of human sexual biology'. *Man*, *21*, 514 (1986).

187. Stoddart, D. Michael. *The Scented Ape: The Biology and Culture of Human Odour*. Cambridge: Cambridge University Press, 1990.

188. Süskind, Patrick. *Perfume: The Story of a Murderer*. New York: Alfred Knopf (1986).

189. Teichmann, H. 'Concerning the power of the olfactory sense of the eel'. *Zeitschrift für vergleichende Physiologie*, *42*, 206 (1959).

190. Tester, A. L. 'The role of olfaction in shark predation'. *Pacific Science*, *17*, 145 (1963).

191. Thiessen, D. D., and M. Rice. 'Mammalian scent gland marking and social behaviour'. *Psychological Bulletin*, *84*, 505 (1976).

192. Thomas, Lewis. *The Lives of a Cell: Notes of a Biology Watcher*. New York: Viking (1974).

193. Twitty, V. C., D. L. Grant and O. Anderson. 'Course and timing of the homing migration of a newt'. *Proceedings of the National Academy of Sciences*, *56*, 864 (1966).

194. Vandenbergh, J. G. 'Male odor accelerates female sexual maturation in mice'. *Endocrinology*, *84*, 658 (1969).

195. Van der Lee, S., and L. M. Boot. 'Spontaneous pseudopregnancy in mice'. *Acta Physiologica et Pharmacologica*, *4*, 442 (1955).

196. Van der Starre, H. (ed.) *Olfaction and Taste VII*. London: IRL Press (1980).

197. Van Hoven, W. 'Trees' secret warning system against browsers'. *Custos*, *13*, 11 (1984).

198. Van Toller, Steve, and G. H. Dodd (eds.). *Perfumery: The Psychology and Biology of Fragrance*. London: Chapman & Hall (1988).

199. Van Toller, Steve, 'Emotion and the brain'. In Ref. 198.

200. Von Frisch, Karl. 'Zur Psychologie das Fisch-Schwarmes'. *Naturwissenschaften*, *26*, 601 (1938).

201. Wallace, P. 'Individual discrimination of humans by odor'. *Physiology and Behavior*, *19*, 577 (1977).

202. Watson, Lyall. *Heaven's Breath: A Natural History of the Wind*. London: Coronet (1984).

203. Watson, Lyall. *Dreams of Dragons: Riddles of Natural History*. London: Sceptre (1987).

204. Watson, Lyall. *Beyond Supernature: A New Natural History of the Supernatural*. New York: Bantam (1988).

205. Weller, Aron, and Leonard Weller. 'Menstrual synchrony between mothers and daughters and between roommates'. *Physiology and Behaviour*, *53*, 943 (1993).

206. Weller, Aron, and Leonard Weller. 'The impact of social interaction factors on menstrual synchrony in the workplace'. *Psychoneuroendocrinology*, *20*, 21 (1995).

207. Wenzel, Bernice. 'The olfactory and related systems in birds'. *Annals of the New York Academy of Sciences*, *519*, 137 (1987).

208. Whitten, W. K. 'Modification of the oestrus cycle of the mouse by external sexual stimuli associated with the male'. *Journal of Endocrinology*, *13*, 399 (1956).

209. Wiener, Harry. 'External chemical messengers: I'. *New York State Journal of Medicine*, *66*, 3153 (1966).

210. Wiener, Harry. 'External chemical messengers: II'. *New York State Journal of Medicine*, *67*, 1144 (1967).

211. Wiener, Harry. 'External chemical messengers: III'. *New York State Journal of Medicine*, *67*, 1286 (1967).

212. Wilkie, Maxine. 'Scent of a market'. *American Demographics*, *17*, 40 (1995).

213. Williams, Joseph M. 'Synesthetic adjectives'. *Language*, *52*, 461 (1976).

214. Wright, Karen. 'The sniff of legend'. *Discover*, April, p. 61 (1994).

215. Wysocki, Charles J., and John J. Lepri. 'Consequences of removing the vomeronasal organ'. *Journal of Steroid Biochemistry and Molecular Biology*, *39*, 661 (1991).

Index

READ MORE IN PENGUIN

In every corner of the world, on every subject under the sun, Penguin represents quality and variety – the very best in publishing today.

For complete information about books available from Penguin – including Puffins, Penguin Classics and Arkana – and how to order them, write to us at the appropriate address below. Please note that for copyright reasons the selection of books varies from country to country.

In the United Kingdom: Please write to *Dept. EP, Penguin Books Ltd, Bath Road, Harmondsworth, West Drayton, Middlesex UB7 0DA*

In the United States: Please write to *Consumer Sales, Penguin Putnam Inc., P.O. Box 12289 Dept. B, Newark, New Jersey 07101-5289*. VISA and MasterCard holders call 1-800-788-6262 to order Penguin titles

In Canada: Please write to *Penguin Books Canada Ltd, 10 Alcorn Avenue, Suite 300, Toronto, Ontario M4V 3B2*

In Australia: Please write to *Penguin Books Australia Ltd, P.O. Box 257, Ringwood, Victoria 3134*

In New Zealand: Please write to *Penguin Books (NZ) Ltd, Private Bag 102902, North Shore Mail Centre, Auckland 10*

In India: Please write to *Penguin Books India Pvt Ltd, 11 Community Centre, Panchsheel Park, New Delhi 110017*

In the Netherlands: Please write to *Penguin Books Netherlands bv, Postbus 3507, NL-1001 AH Amsterdam*

In Germany: Please write to *Penguin Books Deutschland GmbH, Metzlerstrasse 26, 60594 Frankfurt am Main*

In Spain: Please write to *Penguin Books S. A., Bravo Murillo 19, 1° B, 28015 Madrid*

In Italy: Please write to *Penguin Italia s.r.l., Via Benedetto Croce 2, 20094 Corsico, Milano*

In France: Please write to *Penguin France, Le Carré Wilson, 62 rue Benjamin Baillaud, 31500 Toulouse*

In Japan: Please write to *Penguin Books Japan Ltd, Kaneko Building, 2-3-25 Koraku, Bunkyo-Ku, Tokyo 112*

In South Africa: Please write to *Penguin Books South Africa (Pty) Ltd, Private Bag X14, Parkview, 2122 Johannesburg*

READ MORE IN PENGUIN

SCIENCE AND MATHEMATICS

Six Easy Pieces Richard P. Feynman

Drawn from his celebrated and landmark text *Lectures on Physics*, this collection of essays introduces the essentials of physics to the general reader. 'If one book was all that could be passed on to the next generation of scientists it would undoubtedly have to be *Six Easy Pieces*' John Gribbin, *New Scientist*

A Mathematician Reads the Newspapers John Allen Paulos

In this book, John Allen Paulos continues his liberating campaign against mathematical illiteracy. 'Mathematics is all around you. And it's a great defence against the sharks, cowboys and liars who want your vote, your money or your life' Ian Stewart

Dinosaur in a Haystack Stephen Jay Gould

'Today we have many outstanding science writers ... but, whether he is writing about pandas or Jurassic Park, none grabs you so powerfully and personally as Stephen Jay Gould ... he is not merely a pleasure but an education and a chronicler of the times' *Observer*

Does God Play Dice? Ian Stewart

As Ian Stewart shows in this stimulating and accessible account, the key to this unpredictable world can be found in the concept of chaos, one of the most exciting breakthroughs in recent decades. 'A fine introduction to a complex subject' *Daily Telegraph*

About Time Paul Davies

'With his usual clarity and flair, Davies argues that time in the twentieth century is Einstein's time and sets out on a fascinating discussion of why Einstein's can't be the last word on the subject' *Independent on Sunday*